Drinking Water Treatment 1

Drinking Water Treatment 1

Water Quality and Clarification

Kader Gaid

WILEY

First published 2023 in Great Britain and the United States by ISTE Ltd and John Wiley & Sons, Inc.

ISTE Ltd
27-37 St George's Road
London SW19 4EU
UK

www.iste.co.uk

John Wiley & Sons, Inc.
111 River Street
Hoboken, NJ 07030
USA

www.wiley.com

Cover illustration:
© imageBROKER.com/Matton Images

Library of Congress Control Number: 2022941870

British Library Cataloguing-in-Publication Data
A CIP record for this book is available from the British Library
ISBN 978-1-78630-783-5

Contents

Introduction

In the 1990s, microbiological contamination was a major concern, in particular due to the contamination of drinking water sources by cysts and oocysts (*Giardia* and *Cryptosporidium*), as well as by bacteria and viruses. The rise in living standards, the rapid extension of urbanization and ever-increasing industrialization not only contributed to the increase in the demand for water but also to the degradation of receiving environments. In some cases, not even well-protected groundwater was spared. At some point, it became necessary to face the facts and to design an array of technical solutions to revitalize surface water and groundwater to reach their expected properties and qualities in relation to health regulation. On the other hand, the new millennium brought a series of new challenges. Research into the health effects of several inorganic contaminants, such as arsenic and lead, encouraged the reduction of their respective concentrations in water. Lead and water chemistry problems are often related to the corrosion of pipes and other equipment, to such an extent that the focus should not only be on water stability but also on distribution systems. Currently, new pathogens such as new viruses (H_5N_1, H_5N_5, Ebola, Covid-19, etc.) have appeared in our environment, and their treatability in drinking water plants is being tested and monitored. Other concerns are related to the possible health effects of many organic micropollutants. Not only do these include the categories of compounds emanating from our modern way of life – such as pharmaceutical compounds, endocrine-disrupting compounds, and personal care products – but also solid micropollutants, such as microplastics. The appearance of all these physical, organic and biological contaminants in water raises fears of contamination, which may require a more complex treatment. This leads to increased attention on the conditions for protecting water resources, plant performance optimization and a proper management of recycling flows through the implementation of new technologies: high- and low-pressure membrane techniques, a more formalized use of powdered activated carbon, controlled ozonation and ultraviolet radiation. Terrorist attacks place greater emphasis on the security of water infrastructure and risk management, namely, on the transport of dangerous chemicals. Relatedly, regulations are increasingly shifting the point of quality control from factory-produced water to the customer's tap. Considerable media attention has brought these categories of pollutants into the spotlight. It is also true that over the past few years, analytical methods have improved in such a way that trace amounts can now be systematically detected. Even though their toxicity effects on humans, as well as their impact on the environment, are continuous for some of these substances, little is known about their actual effects. The reuse of municipal wastewater as a drinking water resource is also the subject of increased attention and requires the serial use of several high-level techniques.

The microbiological quality of drinking water makes the news every day. At present, customers are increasingly involved in the quality of the water supply and share their concerns about chemical disinfection and its by-products, as well as the presence of lead and legionella in water distribution systems.

For this reason:

– the challenges of drinking water treatment are diverse, making it one of the most interesting fields in civil engineering: new contaminants are under scrutiny, as well as perfluorinated compounds in groundwater and algal toxins that are detectable due to a proliferation of cyanobacteria in surface water;

– alternative resources are being contemplated to profitably recover rainwater, deep brackish groundwater, and fossil water and to directly or indirectly reuse wastewater.

Over the past two decades, there has been a continuous development of new water treatment technologies, and older technologies have also been improved. Thus far, progress has been made in:

– the development and maturation of membrane technologies: microfiltration and ultrafiltration are often used instead of media filtration for particle removal. Nanofiltration and low-pressure reverse osmosis are increasingly used for water softening and the removal of heavy metals, organic carbon and organic micropollutants. High-pressure reverse osmosis is experiencing significant development for desalination applications and for removing organic and mineral contaminants;

– the use of powdered activated carbon (PAC) or micrograin carbon in specific reactors for the removal of humic substances and organic micropollutants;

– the use of biological processes for ordinary applications such as iron and selenium removal from groundwater;

– the development of new and improved advanced oxidation processes for increasing the formation of hydroxyl radicals, therefore inducing the decomposition of many worrisome organic contaminants;

– improvement in the performance of UV disinfection systems to such a point that the currently applied dose is higher than the one required during the 1980s and 1990s.

What about the future?

Nowadays, engineers in charge of designing water treatment processes are confronted with many challenges. In fact, not only are there new challenges, but past challenges are cumulative and still need to be overcome. Some challenges are even contradictory, such as the need for more effective disinfection contrasted with the urge to reduce disinfection by-products. The major challenge is to develop and implement more effective methods for the removal of disinfection by-product (DBP) precursors from raw water or to reduce the risk of protozoa-related contamination without significantly increasing the formation of by-products.

The science of drinking water continues to make progress, and knowledge about the public health risks entailed by emerging contaminants and pathogens is closely monitored.

The biggest challenges in our days are related not only to the treatment processes themselves but also, most importantly, to the choice of treatment route and water supply development.

The water deficit observed in many regions, as well as an increased competition for available water resources, prompts a desire to use every water resource at hand.

There is also an increasing focus on the choice of the most effective water processes. Wastewater is being recycled for various applications, even for drinking water. In fact, in some countries, the supply from groundwater has become common.

In addition, there is an emphasis on the design and supply of more sustainable facilities. Engineers today are called upon to:

– make an optimal use (or reuse) of existing infrastructure and materials;

– limit or reduce the physical and environmental footprint of a facility;

– conserve and recover energy;

– minimize the use of chemicals and the production of solid sewage sludge;

– assess and maybe incorporate sources of alternative energy, in particular those capable of producing a net reduction in greenhouse gas emissions; and

– consider various environmental management benchmarking and accounting procedures, such as the carbon footprint.

Climate change concerns are closely related to water resource challenges and the need to provide sustainable water plants. However, considering that the impacts of climate change are difficult to predict, various questions remain:

– Will the frequency of rainfall be sufficient?

– Will the region see an increase in average temperatures, or will they be more extreme?

– To what extent should the engineer modify treatment routes and make provisions for potential changes, recognizing that these may lead to an increase in the cost of water?

To date, the main discharge-related concerns have focused on the future of sewage sludge, but the environmental impact of liquid discharges through the use of concentrates (high-pressure membrane plants and desalination procedures) and eluates (resulting from water softeners based on ion exchange resins) is now receiving particular attention.

Until recent decades, new water treatment projects and the resulting quality of drinking water have not particularly attracted public attention. Media attention to water issues and the explosion of social media communication are now inviting a larger public engagement in drinking water. As a result, engineers are expected to share certain aspects of a project with the audience in further detail. For this, engineers must take into account a variety of novel requirements related to public perception or values that do not strictly correspond to technical issues. By way of example, this could be the claim that an aluminum residual in the water (even if below standard) could lead to Alzheimer's disease, although there currently is no scientific evidence proving that it constitutes a threat to public health. The audience also expects public services to pay more attention to the esthetic quality of the water delivered. This may include the reduction of taste, odor and color, for example. On the other hand, the reduction of iron, manganese and water hardness is related to comfort parameters rather than potential toxicity effects. In addition to meeting current drinking water standards, engineers are now required to anticipate potential future needs. A water system designed today must be conceived with enough flexibility so that it can be modified to meet potential future requirements. Engineers must also account for other environmental concerns, such as waste management practices, chemical supplies and storage operations, energy conservation, occupational health and safety, and general safety.

Since water treatment engineering is a global market, engineers must adapt to the local context and to the ideas, practices, products and services exchanged. The French water school has always been the subject of particular interest due to its practices, references, technology and services. Building information modeling (BIM) is one element of new practices and technologies that provide the designer, builder and operator with outstanding features, such as:

– the ability to create 3D visualizations throughout the different design stages;

– the possibility of sharing various project aspects between the designer, equipment suppliers, contractors and operators;

– the ability to integrate asset and maintenance management needs.

This book provides a thorough presentation of techniques for producing drinking water, including conventional and unconventional surface water, as well as groundwater. The first chapter presents the diverse physical and chemical constituents most frequently analyzed in drinking water and those requiring reduction. Regardless of whether the resource is surface water (river, lake, dam, ocean, sea) or groundwater, the physicochemical composition of raw water presents many physical, chemical and bacteriological parameters. Contaminants are many and varied. They can be of natural origin, such as humic substances (resulting from plant decomposition), calcium hardness and magnesium, heavy metals (iron, manganese, arsenic, nickel, antimony, etc.). Alternatively, they can be of industrial origin, such as pesticides, nitrates and emerging micropollutants (drug residues, endocrine-disrupting compounds, industrial waste). Contaminants can be of biological origin, with the discharge of wastewater into receiving environments and the concomitant presence of pathogenic germs. Then, there is a presentation of the standards and qualities of drinking water and a discussion about the risks generated by the presence of mineral or organic compounds above standard thresholds. In France, water fit for consumption includes all the water intended for drinking, cooking, food preparation or other domestic purposes, as well as the water used by food companies for manufacturing, processing, preserving or marketing products or substances intended for human consumption, including water-origin ice cream.

All these waters must simultaneously fulfill three conditions:

– they have to be free from any number or concentration of microorganisms and parasites or from any other substances constituting a potential danger to human health; they must comply with the predefined quality values (French Government Resolution of January 11, 2007 on the limits and quality references for raw water and water intended for human consumption, as mentioned in articles R. 1321-2, R. 1321-3, R. 1321-7 and R. 1321-38 from the Public Health Code), which are mandatory values;

– they must satisfy quality references following indicative values (French Government Resolution of January 11, 2007).

Compliance with the quality reference values is assessed at "the point of compliance", that is, the consumer's tap (for water supplied by a distribution network).

To produce drinking water that meets existing standards, Veolia has a large number of tools and technologies displayed on thousands of sites where it operates worldwide. The following chapters further explain the various drinking water production techniques included in a global water treatment system. Each process is accompanied by its fundamental and theoretical underlying principles. All cases include a presentation of the required elements for their design, implementation and operation. These technological elements have been designed to remove particulate pollution through coagulation–flocculation–settling and filtration mechanisms, which all represent ineluctable stages for a drinking water plant treating surface water. Settling based on processes such as Cyclofloc®, Actiflo®, Multiflo® or flotation using Spidlow® or Spidflow® Filter are described in terms of their operating parameters, efficiency and associated chemical reagents. The supplementary treatment of filtration through various materials (sand, anthracite, pumice stone) is detailed with the Filtraflo family, for which the diversity of available techniques (gravity and pressure filters) makes it possible to treat each type of water in accordance with the local context and final water quality goals.

In addition to the removal of particulate pollution, that of dissolved pollution is of great importance for many surface waters. The removal of nitrates has an exclusively dedicated chapter because this compound continues to be relevant in our days. Using either a biological process (Biodenit) or ion exchange resins (Ecodenit), Veolia has developed two processes that have been successfully applied throughout Europe. Particular attention is given to the removal of natural organic matter (humic and fulvic substances) whose residuals in the water produced are known to contribute to the formation of carcinogenic by-products after disinfection. The use of granular activated carbon in Filtraflo GAC, or powdered activated carbon, with the development of many PAC reagents in recent years (Actiflo® Carb, Multiflo® Carb, Opacarb® MG and Opacarb® MF), makes it possible to meet the needs of users for the various organic molecules present in water.

Similarly, an entire chapter is devoted to the removal of emerging organic micropollutants (drug residues, endocrine-disrupting compounds, industrial waste) due to the use of activated carbon (powdered, granular and micrograin) or high-pressure membranes (nanofiltration and reverse osmosis). This chapter illustrates the importance Veolia gives to this type of contaminant and describes their removal processes based on micrograin carbon, such as Opacarb® FL, Filtraflo® Carb or

Contact® Carb. This is done in addition to and/or in combination with other conventional processes for the removal of organic matter.

These fundamental stages of water treatment plants need to be supplemented with other no less important phases providing water with the required calcium-carbon balance to prevent it from being too aggressive (concrete destructuring) or corrosive (alteration of metals). Mechanisms and chemical reactions are properly explained, and many Veolia processes are described by means of saturators, remineralization processes in line or in tanks, or limestone filters (FiltraFlo limestone).

Similarly, chemical decarbonation (Actiflo® Softening, Multiflo® Softening, Saphira®, Actina®), ion exchange or nanofiltration is well detailed, since the problem of excess limestone is an ongoing consumer concern. Metal removal (iron, manganese, arsenic, antimony, selenium, nickel) is presented and described under the light of acquired experience. Low-pressure membranes (microfiltration and ultrafiltration) intended to remove particulate and bacteriological pollution are presented through Veolia's know-how and the many references it has built and/or operates. Because of the expertise and feedback gained from the operations of multiple membrane-based plants, Veolia has built an excellent reputation for its know-how and is often consulted by manufacturers to give an opinion on the membranes being developed.

For high-pressure membranes (nanofiltration, low-pressure reverse osmosis, reverse osmosis), Veolia has a rich and varied database on the use of these membranes in conventional surface water and for the desalination of sea water. Many references are explained in this work, together with the technical justification for their choice.

Chemical disinfection and ultraviolet radiation are detailed in depth as the final stage before the distribution of the water produced, emphasizing the technical recommendations to be implemented to obtain the best disinfection results.

This book is intended as a guide for engineers in charge of the design of drinking water plants, driving them through the different processing goals and helping them with the choice and design of physical, chemical and biological facilities. The book is also intended to aid plant operators by giving them access to the fundamental principles of the processes involved in water treatment plants.

2

Physicochemical and Microbiological Composition of Raw Water

In Europe, to be apt for consumption, water must meet strict quality criteria set by a European directive, which is then transcribed into local law. Water intended for human consumption has to meet nearly 63 parameters, which are defined according to the principle of maximum precaution, to protect consumers whose health is most fragile. These parameters are as follows:

– physicochemical parameters corresponding to water characteristics such as pH, temperature, turbidity, conductivity, color, organic matter, hardness and alkalinity. They concern everything related to the natural structure of water;

– the parameters applicable to undesirable substances refer to substances whose presence in small amounts is tolerated by regulations. These could be, for example, the controlled content of fluorine, nitrates, mineral salts, etc.;

– parameters concerning toxic substances, such as pesticides, and heavy metals such as arsenic, lead or chromium;

– microbiological parameters make it possible to analyze the presence of pathogenic germs (bacteria, viruses, parasites, etc.) to be removed during the treatment.

2.1. Water resources

Water intended for human consumption originates from:

– conventional surface water:

- river water;

- water from lakes, dams and reservoirs;

– non-conventional surface water:

- sea water;

- estuary waters;

– groundwater:

- "water tables" fed by infiltration water (rain, rivers, etc.) and limited by the first impermeable layer encountered in the ground;

- semi-deep aquifers;

- deep aquifers;

- non-conventional groundwater;

- groundwater fed by seawater infiltration.

In specific cases, water resources can originate from wastewater reuse (Singapore, Namibia, United States, etc.).

2.1.1. *Physicochemical parameters*

The composition of surface water varies from season to season. During rainy periods, mineralization tends to be lower and the load of organic matter tends to be higher.

In drier periods, mineralization is greater. Water composition from lakes, dams, etc., varies depending on the depth considered. During a dry period, certain physicochemical parameters are modified: one may observe not only a decrease in dissolved oxygen (DO) and a reduction in nitrates but also an increase in iron and manganese. There can also be algae development when temperatures become warmer.

2.1.1.1. *pH*

pH measures the acidity or basicity of water. The evolution of surface water pH implies a modification in the operating conditions of the water treatment sector: choice of reagents, CO_2 degassing, pre-remineralization, pH adjustment, neutralization, etc.

2.1.1.2. *Temperature*

Surface water temperatures can vary between 0.5°C and 30°C. It is the lowest temperature that determines the design of structures where chemical reactions (such as coagulation-flocculation) or biological treatments will be implemented.

On the other hand, it is the highest temperatures that condition the design of membrane treatment stages, such as nanofiltration or reverse osmosis (low and high pressure).

2.1.1.3. Suspended solids

Suspended matter is composed of minerals (sand, silt, clay, etc.) and living substances (plankton, algae), which act as a support for bacteria, viruses and parasites. Suspended solids (SS) are defined as being decantable in 2 h or retained on a micrometer-pore size filter. In particular, they are generators of color and turbidity.

The SS concentration of surface water can vary between 1 and 500 $mg \cdot L^{-1}$. It is much higher and reaches more than 10,000 $mg \cdot L^{-1}$ in times of floods, heavy rains, dam releases, etc. Many examples in Asia or the Maghreb report concentrations of up to 20,000–50,000 $mg \cdot L^{-1}$.

2.1.1.4. Colloidal particles

Colloidal particles are particles in suspension, with a size between 0.01 and 5 μm. They constitute a SS or a part of them does. They can be removed by coagulation-flocculation, which makes it possible to chemically destabilize them and ensure their settling or filtration on media or through direct filtration on microfiltration or ultrafiltration membranes.

2.1.1.5. Turbidity

Turbidity is related to the concentration of suspended matter (including colloidal matter). Turbidity measurements can be expressed in Jackson Turbidity Units (JTU), Nephelometric Turbidity Units (NTU) or Formazin Turbidity Units (FTU).

1 NTU (Nephelometric Turbidity Unit) equals:

= 1 JTU (Jackson Turbidity Unit)

= 1 FTU (Formazin Turbidity Unit)

= 2.5 $mg \cdot L^{-1}$ of silica

= 1 FNU (Formazin Nephelometric Unit)

Turbidity is an easy-to-measure parameter that provides a proper assessment of variations in the quality of raw water.

In surface water, the majority of particles present are in the size range between 0.1 and 10 μm. Due to their sedimentation characteristics, these particles remain suspended in water. They scatter visible light, and this scattering gives water a murky appearance. In addition, their physical properties enable them to adsorb natural organic matter, metals, bacteria and viruses.

The most important role of turbidity in health is its use as an indicator of the effectiveness of drinking water treatment processes. This is particularly true for the filtration stage, which is responsible for the removal of microbial pathogens that may be present in water. Although there is no accurate relationship between the extent of turbidity reduction and pathogen removal, the trend reflected by field results shows that water with low turbidity contains fewer microorganisms. Turbidity reduction, as well as the removal of particles and pathogens, largely depends on the quality of water at the source and the treatment plant chosen (Figure 2.1).

Turbidity also has various implications for water quality and treatment, depending on the nature of the particles involved. High turbidity readings or fluctuating water quality readings leaving a treatment unit may indicate inadequate water treatment.

Figure 2.1. *Rivers with high turbidity. For a color version of this figure, see www.iste.co.uk/gaid/watertreatment1.zip*

Thus, the removal of turbidity via coagulation–flocculation–settling results in an approximate 90–99% particle removal (colloidal and non-colloidal particles), as well as approximately 1–2 log for bacteria, viruses and parasites aggregated into particles or flocs. Therefore, good turbidity removal facilitates and improves final disinfection.

To better protect public health against microbial contamination, filtration systems strive to achieve the target of 0.5 NFU in Europe and <0.2 NTU in other states, such as the Scandinavian countries, United States, Australia, etc.

2.1.1.6. Conductivity

The total mineralization of water can be represented by measuring its conductivity (micro Siemens per cm or $\mu s \cdot cm^{-1}$). It is a suitable indicator of water quality maintenance for a given mineralization target.

2.1.1.7. Hardness and alkalinity

Hardness represents the total concentration of calcium and magnesium in water. Hard water is a type of water that leaves encrusting deposits. Total hardness (TH) is expressed as the sum of calcium hardness (THCa) and magnesium hardness (THMg). That is:

$$TH = THCa + THMg$$

Alkalinity is represented by the complete alkalimetric title, one of the parameters determining water mineralization. It can be represented by total alkalinity (TALK).

It represents the content of carbonates, hydrogen carbonates and free bases of calcium, magnesium and sodium. For pH values between 6.34 and 10.3, all salts are hydrogen carbonates.

It is used as a reference during remineralization or for pH control during the coagulation or softening/decarbonation stages.

2.1.1.8. Iron and manganese

Iron and manganese are not considered major pollutants, although they affect the organoleptic qualities of the water containing them (appearance, color, metallic taste). Their deposition in water supply networks is responsible for corrosion phenomena, resulting in red or black water coming out of the faucet. In addition, these deposits can act as bacterial beds, which can in turn contaminate water at the consumer's tap.

Although these two elements are mostly found in groundwater, they can also be present in surface water (dams, reservoirs).

2.1.1.8.1. Iron

Iron can appear as dissolved or precipitated, depending on pH and on the oxidation–reduction potential (ORP). Soluble iron can be dissolved (ferrous iron) or complexed with organic substances, such as humic or fulvic acids, or with mineral compounds, such as silicates and phosphates.

Insoluble iron most often appears as ferric hydroxide. In some cases, it can be associated with the SSs present in water. To define an iron removal treatment, it is important to know the different forms in which iron appears.

In broad terms:

– in the case of groundwater, it is soluble iron;

– in the case of surface waters, it may be complexed iron.

2.1.1.8.2. Manganese

Manganese can be found in raw water in three main forms:

– Mn^{2+} soluble form;

– complex form with organic matter;

– Mn^{4+} oxidized form.

Manganese is much more difficult to oxidize than iron and may sometimes require two successive oxidation stages (manganese being removed only after iron).

The presence or absence of iron or manganese in raw water determines the preoxidation stage and choice of reagents (Figure 2.2).

a)   b)

Figure 2.2. *Presence of iron (a) and manganese (b) in distributed water. For a color version of this figure, see www.iste.co.uk/gaid/watertreatment1.zip*

2.1.1.9. *Nitrogen compounds*

Nitrogen is most abundant in water in its mineral form. It appears in three predominant forms:

1) ammonium ions (NH_4^+);

2) Nitrite ions (NO_2^-);

3) Nitrate ions (NO_3^-).

Ammoniacal nitrogen and nitrite ions are present in low concentrations in surface waters, whereas nitrate ions can be found in much higher concentrations. When infiltrated with surface water, groundwater can become loaded with nitrates. Due to the discharge of treated wastewater into rivers and soil leaching, it is possible to find ammonia concentrations higher than 4 $mg \cdot L^{-1}$ and nitrate concentrations above 50 $mg \cdot L^{-1}$.

The nitrogen forms analyzed in water are summarized in Table 2.1.

Nitrogen form	Formula
Ammonia (gas)	NH_3
Ammonium ion	NH_4^+
Nitrite	NO_2^-
Nitrate	NO_3^-
Total mineral nitrogen	$NH_3 + NH_4^+ + NO_2^- + NO_3^-$
Total Kjeldahl nitrogen (TKN)	Organic N $+ NH_3 + NH_4^+$
Organic nitrogen	$TKN - (NH_3 + NH_4^+)$
Total nitrogen (TN)	Organic N $+ NH_3 + NH_4^+ + NO_2^- + NO_3^-$

Table 2.1. *Nitrogen forms present in raw and treated water*

2.1.1.10. *Organic matter (total organic carbon and KMnO$_4$ oxidability)*

Organic matter is formed by organic substances mainly present in surface waters (either in dissolved or particulate form). However, if contaminated by surface water, groundwater can sometimes also contain organic matter. This type of molecule may include fulvic acid, humic acid, peptides or carbohydrates. They are represented by total organic carbon (TOC), whose concentration may vary between 1 and 20 $mg \cdot L^{-1}$ (Table 2.2).

The richer the presence of organic matter in the water, the higher the potential to form organochlorine and organobromine oxidation by-products (trihalomethanes or THMs), which become effective at the disinfection stage. For this reason, it is important to reduce the potential for the formation of THMs in the treatment plant by removing TOC until a residual of less than 2 mg·L^{-1} is obtained.

KMnO$_4$ oxidizability makes it possible to determine the content of oxidizable matter (organic and mineral) by potassium permanganate under correctly defined operating conditions (hot, in an acid medium). The relationship between KMnO$_4$ oxidizability and TOC is often approximately 1.5–1.8 for raw water. It is most often less than 1 for treated water.

Compounds	Origins	Comments
Humic substances (humic and fulvic acids)	Vegetal decomposition	THM, HAA precursors
Volatile organic compounds (THMs, trihalomethanes)	Chlorinated water (raw and treated water)	Chlorination by-products
Haloacetic acids (HAAs)	Chlorinated water (raw and treated water)	Chlorination by-products
Trichlorophenol	Chlorinated water (raw and treated water)	Chlorination by-products
Aldehydes, ketones	Ozonated water	Ozonation by-products
Pesticides	Agriculture	Toxic and/or carcinogenic
Micropollutants (drug residues, endocrine-disrupting compounds, industrial residues)	Pharmaceuticals, veterinary products, industries, etc.	Toxic and/or carcinogenic
Polychlorinated biphenyls (PCBs)	Oil contaminated water	Toxic and bioaccumulative
Polyaromatic hydrocarbons (PAHs)	Petroleum by-products	Low dose carcinogens
Chlorinated solvents	Industries	Potentially carcinogenic

Table 2.2. *Organic compounds analyzed in groundwater and surface water*

2.1.1.11. *Color*

Color is due to the presence of dissolved or colloidal organic matter (in particular humic substances), as well as iron (red color), manganese (black color) or algae (green color). A distinction is made between the apparent overall color of unfiltered water and the true color obtained after filtration, representing the fraction of molecules or ions dissolved in water.

Most of the time, the presence of true color implies the presence of dissolved or colloidal organic matter (notably humic substances), as well as the presence of iron (red color), manganese (black color) or algae (green color). A distinction is made between the apparent color, which is the overall color of the unfiltered water, and the true color, which is obtained after filtration and which represents the fraction of molecules or ions dissolved in the water.

The presence of true color implies, most of the time, the presence of dissolved organic matter of the humic type.

The particulate color is explained by the presence of humic acids and certain metal complexes from humic substances that are poorly soluble at drinking water pH. These substances therefore contribute to water turbidity. In addition, clay-particle and humic substance complexes can be found in the form of dispersed particles in the colloidal state, which also contribute to water turbidity (Figure 2.3).

Figure 2.3. *Different colors of surface water, depending on the concentration of humic substances. For a color version of this figure, see www.iste.co.uk/gaid/watertreatment1.zip*

One color unit (CU) is considered to correspond to 1 mg of platinum in the presence of cobalt (platinum/cobalt [Pt/Co] color scale or Hazen scale).

The color of surface water can vary between 15 and 600 mg·L^{-1} Pt/Co. Drinkable water should have a color of less than 5 mg·L^{-1} Pt/Co, with a maximum of 15 mg·L^{-1} Pt/Co.

2.1.2. Algae (including cyanobacteria and cyanotoxins)

Microalgae are photosynthetic eukaryotic microorganisms that live in a humid environment comprising several thousand species. These unicellular species of a generally autotrophic character fix the surrounding inorganic carbon and transform it into organic matter (conversion of solar energy into chemical energy) at the euphotic zone's level, that is, the surface layer of water traversed by sunlight. Depending on turbidity, this zone can vary in depth (a few centimeters to more than 100 m deep).

They are classified into five groups:

– diatoms or yellow algae (maximum growth in winter and spring);

– cyanophyceae or blue algae (maximum growth in summer and autumn);

– chlorophyceae or green algae (maximum growth in summer and autumn);

– zygophyceae;

– euglenophyceae (characteristic of a eutotrophic environment rich in organic matter).

The presence of major nutrients (nitrogen, phosphorus and carbon), as well as favorable physical conditions (sunshine, transparency, heat), encourage algal growth, a phenomenon called eutrophication. This leads to health-related and organoleptic inconveniences (risk of emission of toxic metabolites, especially in relation to blue algae), as well as technical drawbacks for operating treatment facilities.

The consumption of CO_2 present in water reduces the amount of carbonic acid in it. As a result, a rise in pH is observed, which is then accompanied by algal development.

However, cyanobacteria are of the greatest concern because of the toxins they release. Cyanobacteria is the scientific name given to blue algae floating on the surface of ponds. The first identified species were blue in color, from which the algae took their name. The species identified since then are of various colors, ranging from olive green to red. Cyanobacteria are formed in shallow, warm, calm or still water. Their cells can contain poisons, that is, cyanobacterial toxins. A mass of cyanobacteria in water is called a bloom (Figure 2.4).

Figure 2.4. *Algal (cyanobacterial) bloom in a retainer. For a color version of this figure, see www.iste.co.uk/gaid/watertreatment1.zip*

Algae concentrations in dam water can reach several tens of millions of units per liter (Figure 2.5).

Turbulence and high water flows do not promote the growth of cyanobacteria since they interfere with their ability to maintain a certain position in the water column.

Cyanobacteria species often associated with toxicity are *Anabaena aphanizomenon*, *Microcystis*, *Oscillatoria* and *Nodularia* (Figure 2.6).

Cyanobacteria can release toxins into the water, particularly during cell lysis, which often occurs when they pass through the conventional water purification system. In this sense, cyanobacterial toxins are important contaminants of aquatic ecosystems and constitute a risk for human health. Cyanotoxins may appear in their intracellular (in the cyanobacterium) or extracellular (released in water) form. Cyanobacterial toxins can be classified into different categories: some of them may attack the liver (hepatotoxins) or the nervous system (neurotoxins), whereas others may only irritate the skin.

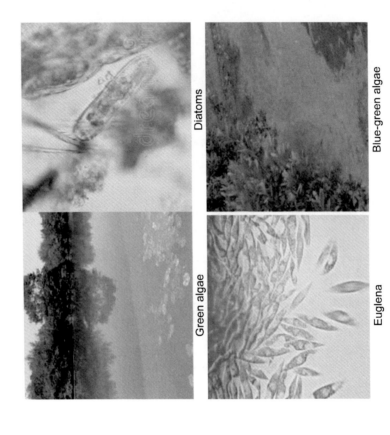

Figure 2.5. *Different types of surface water algae.*
For a color version of this figure, see www.iste.co.uk/gaid/watertreatment1.zip

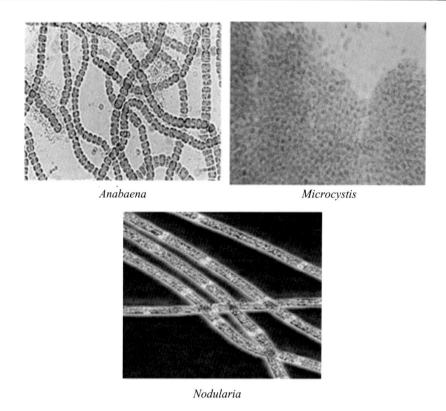

Anabaena   *Microcystis*

Nodularia

Figure 2.6. *Cyanobacteria. For a color version of this figure, see www.iste.co.uk/gaid/watertreatment1.zip*

Hepatotoxins are toxins that attack the liver. Most of the hepatotoxins produced and released by cyanobacteria are called microcystins, because the first hepatotoxin isolated from a cyanobacterium was called *Microcystis aeruginosa*.

Microcystins are the cyanobacterial toxins most frequently found in water and the most common cause of intoxication in animals and humans who come into contact with their poisonous blooms. Microcystins are extremely stable in water due to their chemical structure, which enables them to survive in warm and cold waters and to tolerate significant changes in water chemistry, including water pH. Microcystin-LR is the most commonly found microcystin in water supplies. It has a molecular weight of approximately 1000 Da.

Structurally, microcystins are monocyclic heptapeptides containing two variable L-amino acids and two novel D-amino acids. Microcystins (MC) are named according to their variable L-amino acids (e.g. MC-LR contains leucine (L) and arginine (R), whereas MC-YA contains tyrosine (Y) and alanine (A); see Figure 2.7).

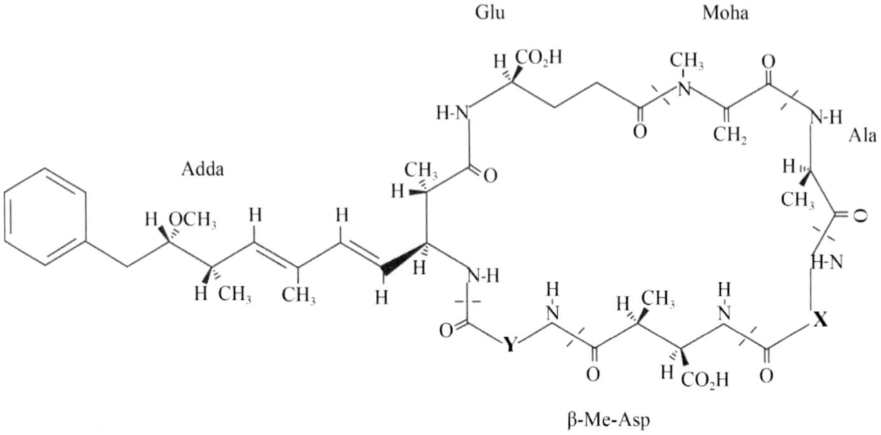

Figure 2.7. *Microcystin structure*

COMMENT ON FIGURE 2.7.–

MC-LR: X = Leucine and Y = Arginine.

MC-LA: X = Leucine and Y = Alanine.

MC-YR: X = Tryptophan and Y = Arginine.

MC-RR: X = Arginine and Y = Arginine.

Neurotoxins (e.g. anatoxin-a) are less prevalent than hepatotoxins in water supplies. It is recommended to screen for the presence of MCs as soon as concentrations of cyanobacteria greater than 20,000 cyanobacteria per milliliter are counted.

2.1.3. Tastes and odors

These two parameters can influence one another. While tastes suggest the presence of mineral constituents, odors suggest the presence of organic substances.

There can be wide variation in taste and odor thresholds among consumers, possibly even exceeding the thresholds usually reported by the panel of judges. Many consumers adapt to them to such an extent that tastes are no longer noticed.

Problems associated with taste particularly arise from the concentration of dissolved salts and the presence of metallic ions such as iron, manganese, copper and zinc. In general, the concentration of salts in drinking water must be less than 1,500 mg·L^{-1}. A concentration of 500 mg·L^{-1} is recommended.

However, consumer complaints and inconveniences arise when these two parameters result from the presence of organic compounds, which can have a varied origin. These compounds can be present in raw water, failing to be removed in the treatment plant (algal toxin, methylisoborneol and geosmin, chlorinated solvents in groundwater, etc., or formed during the disinfection stage due to the action of chlorine. This can be, for example, the odor given off by chloroform (detected from 100 µg·L^{-1}), or the action of chlorine on phenols, thus producing chlorophenols, whose odor is very persistent.

In addition, poorly controlled chloramine disinfection can lead to the presence of di- and trichloramine, which are very strong odorous compounds.

Changes in the quality of water supply networks can lead to the appearance of detectable molecules by taste or odor (Table 2.3).

Odor compounds	Types of odor	Odor thresholds ppm by volume
Amines CH_3NH_2, $(CH_3)NH$	Fish	21
Gaseous ammonia (NH_3)	Ammoniacal	46.8
Diamines $NH_2(CH_2)_4NH_2$, $NH_2(CH_2)_5NH_2$	Rotten flesh	–
Hydrogen sulfide (H_2S)	Rotten eggs	0.00047
(Ethyl and methyl) mercaptans $CH_3(CH_2)SH$, CH_3SH	Rotten cabbage	0.00019, 0.0021.
Organosulphides $(CH_3)_2S$, $(C_6H_5)_2S$	Rotten cabbage	0.0001, 0.0047.
Skatole C_9H_9N	Feces	0.019

Table 2.3. *Odor type and threshold in raw water*

These two parameters are often generated by the presence of algae: bacteria such as actinomycetes (*Streptomyces*, *Nocardia*) and cyanobacteria (*Anabaena*,

Oscillatoria, Phormidium) generate metabolites (methylisoborneol and geosmin), which give off the taste and odor of mold and earth. In fact:

– they constitute the most frequently listed taste and odor problems. Their emergence is most favorable in the spring and summer periods;

– they are blue-green algae, pigmented flagellates, diatoms, etc., whose metabolites can generate the taste and odor of grass, rot, septic tank, fish, cucumber, spices, etc.

2.1.4. *Micropollutants*

2.1.4.1. *Pesticides*

Pesticides are products whose chemical properties contribute to the protection of plants. They are intended to destroy, limit or repel undesirable elements to plant growth, as well as insects, parasites and other plants. They fight against crop diseases or are used for weeding. Pesticides are exclusively of anthropogenic origin. They are often classified depending on the target pest: insecticides (insects), acaricides (mites), aphicides (aphids), ovicides (eggs), larvicides (larvae), herbicides (undesirable plants), fungicides (fungi), molluscicides (molluscs), helicides (snails), rodenticides (rodents), talpicides (moles), corvicides (birds), termicides (termites), repellents, etc.

Surface runoff, drainage and erosion carry these products to surface waters, whereas groundwater can be contaminated by infiltration (transfer through the subsoil). In France, the families of pesticides most often analyzed in groundwater are triazines (atrazine, desethylatrazine, etc.), substituted ureas (diuron) and organochlorines (lindane).

When present in water, pesticides or pesticide residues are considered micropollutants, and their concentration in water is expressed in $\mu g \cdot L^{-1}$. Phytosanitary products constitute a broadly heterogeneous category of substances comprising a large number of molecules (approximately 800).

The toxic mechanisms of pesticides (action on the nervous or respiratory systems, metabolic pathways, etc.) can affect non-target organisms and may be evident in mammals, including humans, resulting in dangerous side effects. Consumers exposed via drinking water and food are prone to this type of toxicity.

Reproductive disorders have often been related to endocrine-disrupting effects exerted on the sexual sphere and have been particularly associated with DDT. These effects extended to other environmental contaminants might be due to their ability to mimic, disrupt or oppose the action of endogenous hormones at various levels (synthesis, transport, activity). The following hypothesis is currently under discussion: very early, or even *in utero*, exposure to "endocrine-disrupting compounds" – and in particular those with estrogenic properties – may entail a wide variety of adverse effects: breast, prostate, testicular cancer, male genitalia malformations (cryptorchidism, hypospadias), endometriosis, male or female infertility and sexual behavior disorders. It should also be noted that these endocrine disorders not only concern sexual physiology but also other hormonal activities (such as thyroid and adrenal functions) and could also affect the immune system.

Table 2.4 shows a list of 44 pesticides suspected of being endocrine-disrupting compounds (EDCs).

2,4,5-T DBCP	DBCP	H-Epoxide	Nitrofen
2,4-D	DDT	Kelthane	Oxychlordane
Alachlor	DDT Metabolites	Kepone	Permethrin
Aldicarb	Dicofol	Malathion	Pyrethroids
Amitrole	Dieldrin	Mancozeb	Toxaphene
Atrazine	Endosulfan	Maneb	Transnonachlor
Benomyl oxide	Esfenvalerate	Methomyl	Tributyltin
Beta-HCH	Ethylparathion	Methoxychlor	Trifluralin
Carbaryl	Fenvalerate	Metiram	Vinclozolin
Chlordane	Lindane	Metribuzin	Zineb
Cypermethrin	Heptachlor	Mirex	Ziram

Table 2.4. *List of pesticides suspected of being endocrine-disrupting compounds*

The French government resolution of January 11, 2007, sets a quality requirement of 0.1 $\mu g \cdot L^{-1}$ for phytosanitary products per individual substance (except four of them: aldrin, dieldrin, heptachlor and heptachlor epoxide, for which the applicable limit is 0.03 $\mu g \cdot L^{-1}$, following the WHO guideline value) and 0.5 $\mu g \cdot L^{-1}$ for the total amount of pesticides tolerated.

Other legislation recommends a standard value for each phytosanitary product.

2.1.4.2. *Pharmaceutical residues, industrial products and endocrine-disrupting compounds*

The term *micropollutant* designates natural or synthetic-origin substances, likely to have a toxic action at very low concentrations (in the range of $\mu g \cdot L^{-1}$), in a given environment. Some of these substances are qualified as EDCs because they can act at very low doses on the hormonal balance of living beings.

Thus, according to the European Commission, *"An endocrine disruptor is an exogenous substance or mixture that alters function(s) of the endocrine system and consequently causes adverse health effects in an intact organism, or its progeny, or (sub)populations."*

Due to the diversity and complexity of the substances considered, data are still limited on the effects of pharmaceutical residues (PRs) and EDCs on the environment and their occurrence patterns in water resources. Due to the wide diversity of the molecules considered and the difficulties related to their analysis – which have sometimes been associated with their low concentrations – the studies carried out until now to assess the presence of these compounds in the environment have been limited to a few groups of molecules. These were chosen on the basis of production amounts and their estimated persistence in the environment (for PRs), as well as the potential endocrine activity for EDCs.

It should be noted that the types and concentrations of micropollutants in wastewater vary greatly from one wastewater treatment plant to another: they depend on the activity around the station (domestic, industrial, agricultural), as well as its setting (e.g. soil pollution during rainfall runoff).

Conventional wastewater treatment plants already treat some micropollutants present in wastewater. However, to obtain an efficient treatment of micropollutants, it is necessary to add specific treatment stages to the process.

The European Commission has published a list of priority substances to monitor in surface waters, published in September 2015, including 45 numbered micropollutants, namely, heavy metals, pesticides, aromatic hydrocarbons, chlorine compounds, etc. (Table 2.5).

In June 2018, a "watch list" was added to the list of priority substances, whose purpose is to acquire further knowledge about certain substances that could potentially be included under the list of priority substances in the future (Table 2.6).

1. Alachlor	20. Lead and its compounds	35. Perfluorooctane sulfonic acid and its derivatives (perfluoro-octanesulfonate PFOS)
2. Anthracene	21. Mercury and its compounds	
3. Atrazine		
4. Benzene	22. Naphthalene	36. Quinoxifen
5. Only brominated diphenyl ethers:	23. Nickel	37. Dioxins and dioxin-like compounds (1)
Tetrabromodiphenyl ether,	24. Nonylphenols including 4-nonylphenol and 4-nonylphenol (branched) isomers	38. Aclonifen
pentabromodiphenyl ether,		39. Bifenox
hexabromodiphenyl ether,	25. Octylphenols including 4-(1,1',3,3'-tetramethylbutyl)-phenol isomer	40. Cybutryne
Heptabromodiphenyl ether		41. Cypermethrin (2)
6. Cadmium and its compounds		42. Dichlorvos
7. Chloroalkanes, C10-13	26. Pentachlorobenzene	43. Hexabromocyclododecane (HBCDD) (3)
8. Chlorfenvinphos	27. Pentachlorophenol	
9. Chlorpyrifos (ethylchlorpyrifos)	28. Hydrocarbons:	44. Heptachlor and heptachlor epoxide
10. 1,2-Dichloroethane	Polycyclic aromatic hydrocarbons (PAHs) including benzo[a]pyrene, benzo[b]fluoranthene, benzo[ghi]perylene, benzo[k]fluoranthene, and indeno[1,2,3-cd] pyrene, but excluding anthracene, fluoranthene, and naphthalene, which are listed separately.	45. Terbutryne
11. Dichloromethane		
12. Di(2-ethylhexyl) phtalate (DEHP)		
13. Diuron		
14. Endosulfan		
15. Fluoranthene		
16. Hexachlorobenzene		
17. Hexachlorocyclohexane	29. Simazine	
18. Hexachlorobutadiene	30. Tributyltin compounds	
19. Isoproturon	31. Trichlorobenzene	
	32. Trichloromethane (Chloroform)	
	33. Trifluralin	
	34. Dicofol	

Table 2.5. *List of 45 priority substances (French government resolution of September 7, 2015)*

NOTES ON TABLE 2.5.–

1) Refers to the following compounds:

– seven polychlorinated dibenzo-p-dioxins (PCDD): 2,3,7,8-T4CDD, 1,2,3,7,8-P5CDD, 1,2,3,4,7,8-H6CDD, 1,2,3,6,7,8-H6CDD, 1,2,3,7,8,9-H6CDD, 1,2,3,4,6,7,8-H7CDD, 1,2,3,4,6,7,8,9-O8CDD;

– ten polychlorinated dibenzofurans (PCDF): 2,3,7,8-T4CDF, 1,2,3,7,8-P5CDF, 2,3,4,7,8-P5CDF, 1,2,3,4,7,8-H6CDF, 1, 2,3,6,7,8-H6CDF, 1,2,3,7,8,9-H6CDF, 2,3,4,6,7,8-H6CDF, 1,2,3,4,6, 7,8-H7CDF, 1,2,3,4,7,8,9-H7CDF, 1,2,3,4,6, 7,8,9-O8CDF;

– twelve dioxin-like polychlorinated biphenyls (DL-PCBs): 3.3', 4.4'-T4CB (PCB 77, 3.3', 4', 5-T4CB (PCB 81), 2,3.3', 4.4'-P5CB (PCB 105), 2,3,4,4', 5-P5CB (PCB 114), 2,3', 4,4', 5-P5CB (PCB 118), 2,3', 4,4', 5'-P5CB (PCB 123), 3.3', 4.4', 5-P5CB (PCB 126), 2,3,3', 4.4', 5-H6CB (PCB 156), 2,3,3', 4, 4', 5'-H6CB (PCB 157), 2.3', 4.4', 5.5'-H6CB (PCB 167), 3.3', 4.4', 5.5'-H6CB (PCB 169), 2,3,3', 4,4', 5,5'-H7CB (PCB 189).

2) Mixture of isomers from cypermethrin, alpha-cypermethrin, beta-cypermethrin, theta-cypermethrin and zeta-cypermethrin.

3) Refers to 1,3,5,7,9,11-hexabromocyclododecane, 1,2,5,6,9,10-hexabromocyclododecane, α-hexabromocyclododecane, β-hexabromocyclododecane and γ-hexabromocyclododecane.

17-Alpha-ethinyl estradiol (EE2)
17-Beta-estradiol (E2), Estrone (E1)
Macrolide antibiotics
Erythromycin, clarithromycin, azithromycin
Methiocarb
Neonicotinoids
Imidacloprid, thiacloprid, thiamethoxam, clothianidin, acetamiprid
Metaflumizone
Amoxicillin
Ciprofloxacin

Table 2.6. *Watch list (EU 2018/840 of June 5, 2018)*

2.2. Microbiology

Water is loaded with microorganisms due to both the discharge of treated wastewater as well as contact with the ground. These microorganisms are viruses, bacteria and parasites. The bacterial quality of water is assessed after searching for and identifying test germs for fecal contamination.

Potentially pathogenic waterborne microorganisms include bacteria, viruses and parasites.

2.2.1. Bacteria

Contamination by human or animal feces is the main source of pathogens in drinking water. Most waterborne potentially pathogenic bacteria infect the gastrointestinal tract.

Bacteria are single-celled organisms lacking properly defined nuclear membranes. The most common pathogenic bacteria found in water are *Escherichia coli*, *Salmonella*, *Shigella*, *Yersinia enterocolitica*, *Legionella*, *Vibrio cholerae* and *Mycobacterium*.

Total coliforms are enterobacteria commonly found in the environment, for example, in soil or vegetation, as well as in the intestines of mammals and humans. This bacterial group is used as an indicator of the microbial quality of water because it contains fecal-origin bacteria, such as *E. coli*. In summary, total coliforms are mainly useful as indicators of treatment efficacy, distribution network integrity and bacterial regrowth after treatment.

Fecal coliforms, or thermotolerant coliforms, are a subgroup of total coliforms. The species most frequently associated with this bacterial group is *E. coli* and, to a lesser extent, some species of the genera *Citrobacter*, *Enterobacter* and *Klebsiella*.

Escherichia coli belongs to the group of thermotolerant coliforms from the enteric bacteria family. *Escherichia coli* can be of human or animal fecal origin and does not exist in the natural environment. Therefore, the detection of *Escherichia coli* in drinking water is an indicator of fecal contamination, which should lead to serious suspicion about the presence of other pathogenic microorganisms. *Escherichia coli* is the specific indicator identifying this type of contamination, and it is easier to recognize in comparison with other indicators or other specific pathogenic microorganisms (Figure 2.8).

In general, the detection of fecal coliforms in treated water can be an indicator of the presence of entero-pathogenic microorganisms, such as salmonella and Norwalk virus.

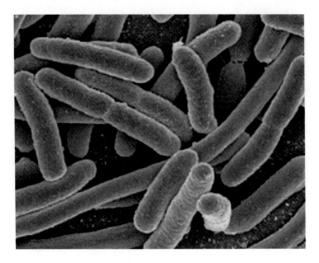

Figure 2.8. *Escherichia coli*

However, some pathogenic waterborne bacteria, such as legionella, can grow both in water and soil. These bacteria are mainly transmitted by inhalation and by contact (toilets) and can infect the respiratory tract or the brain.

The genus *Enterococcus* includes approximately 20 species found in various habitats and different hosts. They are often found in the gastrointestinal tract of humans and several animals; *Enterococcus faecalis* and *Enterococcus faecium* are the two most frequently identified species in humans.

The persistence of enterococci in various types of water can be higher than that in other indicator organisms, particularly due to their resistance to disinfectants, which makes them the preferred indicators for evaluating the effectiveness of water treatment. Since there is typically no enterococci growth in distribution networks, their detection is generally attributed to fecal pollution. The role of enterococci has recently been recognized as an indicator of fecal contamination in aquifers (groundwater).

2.2.2. Viruses

Viruses are extremely small microorganisms lacking the capacity to reproduce outside of a host cell. A virus is an infectious parasitic microbe, almost entirely made up of proteins and nucleosides.

In general, viruses are host specific and can replicate only in living cells. Their size can vary between 0.020 and 0.35 µm in diameter. Viruses that can replicate in the gastrointestinal tract of humans or animals are called "enteric viruses".

Enteric viruses are excreted in the feces of infected people or animals, and some enteric viruses can also be excreted in the urine.

In addition to gastroenteritis, enteric viruses can cause serious acute illnesses, such as meningitis, poliomyelitis and non-specific febrile illnesses. They have also been involved in the etiology of certain chronic diseases, such as diabetes mellitus and chronic fatigue syndrome. Enteric viruses commonly associated with human waterborne disease include noroviruses, hepatitis A virus (HAV), hepatitis E virus (HEV), rotaviruses and enteroviruses (Figure 2.9).

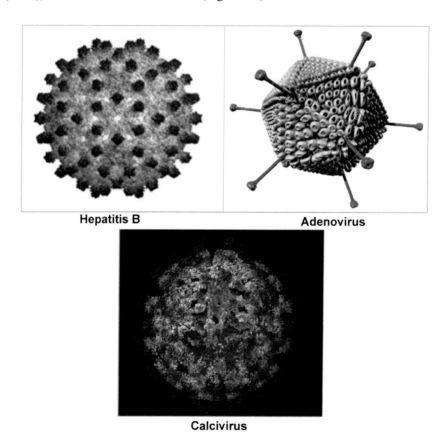

Hepatitis B   **Adenovirus**

Calcivirus

Figure 2.9. *Viruses identified in water. For a color version of this figure, see www.iste.co.uk/gaid/watertreatment1.zip*

They seem to be highly present in surface waters. In the case of groundwater, viruses have been detected both in confined and unconfined aquifers. The absence of indicator bacteria (e.g. *E. coli*, total coliforms) does not necessarily mean that there are no enteric viruses.

In general, viruses are more persistent in the environment and more resistant to disinfection treatments than bacteria. A minimum 4-log removal and/or inactivation of enteric viruses is recommended for all water sources, including groundwater sources. For many water sources, a greater than 4-log reduction may be necessary.

It is important to note that new enteric viruses continue to be detected and recognized.

2.2.3. Parasites (Cryptosporidium and Giardia)

Parasites may be present in water resources contaminated by sewage discharges (fecal discharges from humans and animals). They are also found in fruits and vegetables grown or washed with contaminated water.

The species that is almost exclusively responsible for human infection is *Cryptosporidium parvum*. The lifecycle of *Cryptosporidium* comprises six stages. The cycle begins with the ingestion of oocysts (4–6 μm in diameter) by the host, which then undergo decysting, thus releasing sporozoites that parasitize the gastrointestinal epithelial cells. The sporozoites mature into trophozoites and then into merozoites, which infect other epithelial cells (this stage, called merogony, corresponds to asexual reproduction). Merozoites initiate sexual reproduction by giving rise to gametes, which eventually develop into oocysts.

Approximately 20% of oocysts are thin-walled and are responsible for maintaining the infection in the host; however, the majority of oocysts (approximately 80%) develop a thick double wall and are removed through feces, thus contaminating the environment. It is also important to observe that the correlation between the number of oocysts and microbial indicators – such as total coliforms and fecal coliforms – is relatively weak or non-existent. In this context, the various microbial indicators could not be used as a useful source of information to detect an increased concentration of parasites in the water. This explains why the persistence of low and constant turbidity constitutes the best indicator that the presence of *Cryptosporidium* is unlikely, not even in small amounts.

Giardia lamblia is a protozoan flagellate required to parasitize a host to complete its lifecycle, which includes two forms:

– the non-infectious, mobile trophozoite (with several flagella), which cannot survive outside the host due to its fragility;

– the cyst (approximately 8 × 14 μm), which is the infectious form and can survive in various adverse environmental conditions.

The lifecycle generally begins with the infection of the host following the ingestion of cysts present in contaminated food or water. After decystation in the duodenum, two trophozoites are released that attach themselves to the intestinal villi; the trophozoite is able to reproduce asexually, giving rise to two other trophozoites. By migrating toward the colon and under the effect of bile salts, the trophozoite undergoes significant structural and physiological changes that lead to the formation of cysts. The latter are discharged in the environment, where they prove to be very resistant.

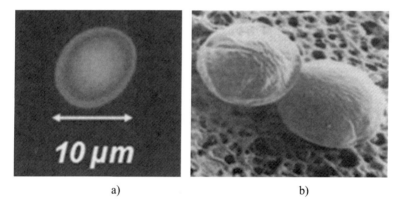

a)   b)

Figure 2.10. *Giardia lamblia (a) and Cryptosporidium (b). For a color version of this figure, see www.iste.co.uk/gaid/watertreatment1.zip*

Groundwater can also be contaminated by *Cryptosporidium* and *Giardia*, especially if it is under the direct influence of surface water (Figure 2.10).

A good correlation between the number of *Giardia* cysts and microbial pollution indicators (total and fecal coliforms), as well as turbidity, has been found in surface waters.

French regulations (January 11, 2007) relating to the quality of drinking water specify that water intended for human consumption must be free from pathogenic organisms, including parasites.

2.3. Quality of water intended for human consumption

All water intended for human consumption must comply with the strictest standards, established for all the countries of Europe, for a large number of microbiological, physical and chemical parameters whose limit values should not be exceeded.

In France, the government resolution of January 11, 2007 states that water intended for human consumption must meet approximately 70 quality criteria divided between quality limits and quality reference values.

Water quality limits are imperative because they can have an impact on health and concern microbiological and chemical parameters. The bacteriological quality of drinking water must be ensured under all circumstances, and no anomalies can be tolerated.

Quality reference values are indicators that reflect the proper functioning of drinking water production facilities and include microbiological, chemical, organoleptic parameters, as well as radioactivity indicators. Their non-compliance may be a sign of a malfunction in the treatment or distribution facilities and should alert the operator.

Screening procedures include the following:

– indicators of the presence of pathogenic microorganisms (bacteria, viruses, parasites) that must not be present in the water, as well as bacterial indicators representative of the overall good quality of distributed water;

– certain chemical compounds considered undesirable or toxic, such as hydrocarbons and pesticides, heavy metals or nitrates;

– certain physicochemical or organoleptic parameters: water turbidity must be as low as possible, water should not have a strong odor or taste, should not be aggressive or corrosive, etc.

The European Directive 98/83/EC of November 3, 1998, relating to the quality of water intended for human consumption, constitutes the European regulatory framework for drinking water. It applies to all waters intended for human consumption, with the exception of natural mineral water and medicinal waters. Table 2.7 summarizes the parametric values of the European Directive and French regulations.

2.3.1. Microbiological parameters

The proper microbiological quality of the water supply is essential to avoid pathologies such as gastroenteritis. In the field of drinking water, microbiological risk represents short-term risk. For this reason, the quality limits of microbiological parameters are subject to zero tolerance.

In French regulations, water potability limits are imperative because they can have an impact on health and concern microbiological and chemical parameters. The water distributed must be free of *E. coli* and *Enterococcus*. The bacteriological quality of drinking water must be ensured in all circumstances and cannot be subject to any tolerance.

Quality references are indicators that reflect the proper functioning of drinking water production facilities and include microbiological, chemical, organoleptic parameters and radioactivity indicators. Their non-compliance may be a sign of a malfunction in the treatment or distribution facilities and should alert the operator. Their analysis takes into account the possible risks for the health of people but also the approval of the use of water for the users. The organoleptic parameters relate to the color, flavor and transparency of the water and have no direct health value.

Contaminants	European Directive 98/83/EC November 3, 1998		French regulations (French Government Resolution January 11, 2007)	
	Parametric value	Indicator parameter	Quality limit	Quality reference value
Escherichia coli	0/100 mL	-	0/100 mL	-
Enterococci	0/100 mL	-	0/100 mL	-
Total coliforms	-	0/100 mL	-	0/100 mL
Sulfite-reducing bacteria, including spores	-	-	-	0/100 mL
Clostridium Perfingens including spores	-	0/100 mL	-	-
Colony content at 22°C	-	No abnormal change	-	-

Contaminants	European Directive 98/83/EC November 3, 1998		French regulations (Government Resolution January 11, 2007)	
	Parametric value	Indicator parameter	Quality limit	Quality reference value
Physicochemical parameters				
Acrylamide ($\mu g \cdot L^{-1}$)	0.1	-	0.10	-
Total Aluminum ($mg \cdot L^{-1}$)	-	0.20	-	0.20
Ammonium (NH_4^+) ($mg \cdot L^{-1}$)	-	0.50	-	0.10
Antimony ($\mu g \cdot L^{-1}$)	5.0	-	5.0	-
Arsenic ($\mu g \cdot L^{-1}$)	10	-	10	-
Barium ($mg \cdot L^{-1}$)	-	-	—	0.70
Benzene ($\mu g \cdot L^{-1}$)	1.0	-	1.0	-
Benzo[a]pyrene ($\mu g \cdot L^{-1}$)	0.010	-	0.010	-
Boron ($mg \cdot L^{-1}$)	1.0	-	1.0	-
Bromates ($\mu g \cdot L^{-1}$)	10	-	10	-
Cadmium ($\mu g \cdot L^{-1}$)	5.0	-	5.0	-
Total Organic Carbon ($mg \cdot L^{-1}$)	-	No abnormal change	-	2.0
Chlorites ($mg \cdot L^{-1}$)	-	-	-	0.20
Chlorites ($mg \cdot L^{-1}$)	-	250	-	250
Vinyl chloride ($\mu g \cdot L^{-1}$)	0.5	-	0.5	-

Contaminants	European Directive 98/83/EC November 3, 1998		French regulations (Government Resolution January 11, 2007)	
	Parametric value	Indicator parameter	Quality limit	Quality reference value
Chromium (µg·L⁻¹)	50	-	50	-
Conductivity (µS·cm⁻¹)	-	<2,500	-	≥180 and ≤1,000 at 20°C ≥200 and ≤1,100 at 25°C
Color (Pt/Co scale)		Acceptable		<15
Copper (mg·L⁻¹)	2.0	-	2.0	-
Total cyanides (µg·L⁻¹)	50	-	50	-
1,2-dichloroethane (µg·L⁻¹)	3.0	-	3.0	-
Epichlorohydrin (µg·L⁻¹)	0.10	-	0.10	-
Calco-carbonic equilibrium	-	-	-	The water must be in calco-carbonic equilibrium
Iron (µg·L⁻¹)	-	200	-	200
Fluorides (mg·L⁻¹)	1.50		1.50	
Polycyclic aromatic hydrocarbons (PAHs) (µg·L⁻¹)	0.10	-	0.10	-
Manganese (µg·L⁻¹)	-	50	-	50
Mercury (µg·L⁻¹)	1.0	-	1.0	-
Total microcystins (µg·L⁻¹)	1.0	-	1.0	-

Contaminants	European Directive 98/83/EC November 3, 1998		French regulations (Government Resolution January 11, 2007)	
	Parametric value	Indicator parameter	Quality limit	Quality reference value
Nickel (μg·L⁻¹)	20	-	20	-
Nitrates (NO₃⁻) (mg·L⁻¹)	50	-	50	-
Nitrites (NO₂⁻) (mg·L⁻¹)	0.50	-	0.50	-
Odor (dilution rate at 20°C)	-	Acceptable	-	<3
KMnO₄ oxidizability (acid medium) (mg·L⁻¹)	-	5.0	-	5.0
Pesticides (per individual substance) (μg·L⁻¹)	0.10	-	0.10	-
Aldrin, dieldrin, heptachlor, heptachlorepoxide (per individual substance) (μg·L⁻¹)	0.03	-	0.03	-
Total pesticides (μg·L⁻¹)	0.50	-	0.50	-
pH (at 20°C)	-	≥6.5 and ≤9	-	≥6.5 and ≤9
Lead (μg·L⁻¹)	10		10	
Selenium (μg·L⁻¹)	10		10	
Sodium (mg·L⁻¹)	-	200	-	200
Sulfates (mg·L⁻¹)	-	250	-	250

Contaminants	European Directive 98/83/EC November 3, 1998		French regulations (Government Resolution January 11, 2007)	
	Parametric value	Indicator parameter	Quality limit	Quality reference value
Temperature (°C)	-	-	-	25 Not applicable in French overseas departments and territories
Tetrachlorethylene and trichloroethylene ($\mu g \cdot L^{-1}$)	-	-	10	-
Total trihalomethanes (THMs) ($\mu g \cdot L^{-1}$)	-	-	100	-
Turbidity at the point of distribution	-	Acceptable and no abnormal change	1.0 NTU	0.5 NFU The quality reference value is applicable at the point of distribution, for the waters referred to under Article R. 1321-37 and for groundwater coming from fractured porous media presenting a significant periodic turbidity, greater than 2.0 NFU at the consumer's tap

| Contaminants | European Directive 98/83/EC November 3, 1998 | | | French regulations (Government Resolution January 11, 2007) |
	Parametric value	Indicator parameter	Quality limit	Quality reference value
Radioactivity parameters				
Overall alpha activity	-	-	-	In the event of a value greater than 0.10 Bq/L, the specific radionuclides are analyzed in compliance with the French Government Resolution mentioned under Article R. 1321-20
Residual global beta activity	-	-	-	In the event of a value greater than 1.0 Bq/L, the specific radionuclides are analyzed in compliance with the French Government Resolution mentioned under Article R. 1321-20
Indicative dose (ID) mSv/year (micro Sieverts per year)	0.10	0.10	-	The calculation of the ID is carried out according to the methods defined under Article R. 1321-20
Radon (Bq·L^{-1})	-	-	-	100 Only for groundwater
Tritium (Bq·L^{-1})	-	100	-	100 The presence of high concentrations of tritium in water may indicate the presence of other artificial radionuclides. In the event of a value greater than the quality reference value, the specific radionuclides are analyzed in compliance with the French Government Resolution mentioned under Article R. 1321-20

Table 2.7. Parametric values of the European Directive and French regulations

2.4. References

Baylar, A., Bagatur, T., Emiroglu, M.E. (2007). Aeration efficiency with nappe flow over stepped cascades. *Proceedings of the Institution of Civil Engineers – Water Management*, 160(1), 43–50.

Baylar, A., Kisi, O., Emiroğlu, M.E. (2008). Aeration efficiency estimation in stepped cascade aerators using neutral network approach. *Natural and Applied Sciences Civil Engineering*, 3(2), 361–371.

Bilello, L.J. and Singley, J.E. (1986). Removing trihalomethanes by packed columns and diffused aeration. *Journal AWWA*, 78(2), 62–71.

Brown, L.C. and Baillod, C.R. (1982). Modeling and interpreting oxygen transfer data. *Jour. Environ. Eng. Div. Proc. Am. Soc. Civ. Eng.*, 108(EE4), 607–616.

Chanson, H. (1994). Hydraulics of nappe flow regime above stepped chutes and spillways. *Aust. Civil Engineering Trans.*, CE36(1), 69–76.

Djebbar, Y. and Narbaitz, R.M. (1995). Mass transfer correlations for air stripping towers. *Environmental Progress*, 14(3), 137–145.

Gross, R.L. and TerMaath, S.G. (1985). Packed tower aeration strips trichloethene from groundwater. *Environmental Progress*, 4(2), 119–124.

Gulliver, J.S., Thene, J.R., Rindels, A.J. (1990). Indexing gas transfer in self-aerated flows. *Journal of Environmental Engineering*, 116(3), 503–523.

Kavanaugh, M.C. and Trussell, R.R. (1980). Design of aeration towers to strip volatile coumpunds from drinking water. *Journal AWWA*, 72(12), 684–691.

Little, J.C. and Selleck, R.E. (1991). Evaluating the performance of two plastic packings in a cross-flow aeration tower. *Journal AWWA*, 83(6), 88–95.

Nakasone, H. (1987). Study of aeration at weirs and cascades. *ASCE Journal of Environmental Engineering*, 113(1), 64–75.

Roberts, P.V., Munz, C., Dändliker, P.G. (1984). Modeling volatile organic solute removal by surface and bubble aeration. *Journal WPCF*, 56(2), 157–163.

Tatewar, S.P. and Ingle, R.N. (1999). Nappe flow on inclined stepped spillways. *Journal of the Institution of Engineers (India)*, 79, 175–179.

Tebbutt, T.H.Y. (1972). Some studies on aeration in cascades. *Water Res.*, 6, 412–425.

Toombes, L. and Chanson, H. (2000). Air–water flow and gas transfer at aeration cascades: A comparative study of smooth and stepped chutes. *Proceedings of the International Workshop on Hydraulics of Stepped Spillways*, 22–24 March, Zurich.

Worm, G.I.M., Mesman, G.A.M., van Schagen, K.M., Borger, K.J., Rietveld, L.C. (2009). Hydraulic modelling of drinking water treatment plant operations. *Drink. Water Eng. Sci.*, 2, 15–20.

3

Aeration and Stripping

The oxygen concentration in surface waters is an important indicator of water quality. Water aeration or reaeration can be an essential stage in the treatment plant if the concentration of dissolved oxygen (DO) is not sufficient. This can be due to a deficit in raw water or because a part of the DO was consumed by chemical or biological reactions throughout the various processing stages.

3.1. Cascade aeration

The simplest aeration or reaeration process is represented by cascade aeration. With average efficiency and generating a sufficient charge drop, cascades constitute an economical means of aeration in terms of investment, which should be reserved for small and medium units. The vertical gap of the upstream/downstream filters or possible repumping should be considered.

A cascade is set up on raw water or after a treatment stage in which DO is consumed.

The cascade aerator is used for in-stream reaeration and is widely used in water treatment plants to enhance the air–water transfer of atmospheric gasses and the removal of iron and manganese, which can cause unpleasant taste and odors.

Thus, the cascade can be proposed:

– when the raw water's DO concentration is low;

– after oxidation of the dissolved iron stage; and

– after biological anoxia treatment, such as biological denitrification or biological iron removal.

It is therefore a re-aeration step:

– for the removal of volatile organic compounds (chloroform and others);

– for the reduction of carbon dioxide;

– to enrich water with O_2 for the biological treatment of ammonia;

– for the removal of gases such as radon, hydrogen sulfide, methane, etc.;

– to produce water containing at least a 75% saturation of DO and to avoid anoxic conditions in distribution networks.

The cascade is an interesting and effective stage for many applications, which should not be confused with stripping (see section 3.4).

The efficiency of the cascade treatment essentially depends on the available geometric height. In practice, there is general consensus not to reasonably go beyond 70–75% saturation with a cascade height of less than 2.70–3.00 m.

3.1.1. *Characteristics*

The basic principle of a cascade aerator is to generate aeration, which is extremely helpful for facilitating the transfer of gas into or out of water. Aeration of water by a stepped cascade system continues to be one of simplest methods.

Cascades are built as openwork steps on a vertical wall. The distribution on successive cascades is performed using spillways (depth of water lower than 10 cm).

The performance of such a system depends on the selected geometry of the cascade system for particular hydraulic loading.

The design takes into account:

– the initial O_2 content;

– the desired final O_2 content;

– the total available height (between the upstream water body and the downstream water body);

– with a slope of approximately 45°;

– with a minimum drop height of 20 cm per cascade;

– downstream isolation, which is generally obtained using wall valves.

3.2. Operating principle of a cascade aerator system

Aeration is a unitary process in which air and water are brought into intimate contact. A cascade aerator makes water fall throughout a lateral sequence of steps located at different levels (called steps or floors), enabling water to be enriched with DO during its fall. The physical process of aeration is based on oxygen transfer from the atmosphere to the water to be treated, with a view to its enrichment. During this treatment stage, oxidation reactions can occur if the physicochemical conditions are met.

The gas–liquid transfer takes place via small air bubbles that are injected all along the water's trajectory. This transfer is accelerated by the strong turbulence of the mixture created by the operating conditions. Turbulence favors an increase in the air/water ratio, inducing oxygen dissolution, provided that sufficient contact time is respected. When water falls from one step to another, an interface between air and water is created. Significant amounts of air are entrained, and the air is then dispersed as bubbles in the body of water. This leads to a gas–liquid transfer that takes place at the interface between the water and air bubbles. Thus, the different steps act as a series of small O_2 transfer reactors, favoring the passage of overflowing water from one step to another (Figure 3.1).

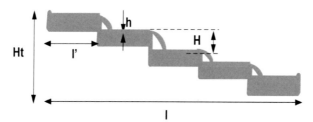

Figure 3.1. *Cross-sectional side view of a stepped cascade. For a color version of this figure, see www.iste.co.uk/gaid/watertreatment1.zip*

It is important for water to fall from one step to another following a homogeneous flow so that the physical transfer conditions are identical at each step's level. The water sheet height (h) above each step is calculated so that the O_2 transfer within this water sheet is optimal and prevents the formation of whirlwinds (which could create disparities and bypasses in the water circuit). Step height (H) measures between 0.3 and 0.5 m, and step width (l′) measures between 0.4 and 0.6 m. The cascade's total width (l) can vary between 1.8 and 6 m, depending on the quality of raw water and treatment goals. The large surface of water thus created favors simple and fast aeration (Figure 3.2).

Figure 3.2. *Stepped cascade at a drinking water plant in India (Veolia). For a color version of this figure, see www.iste.co.uk/gaid/watertreatment1.zip*

The rate of oxygen mass transfer from the gas (air bubbles) to the liquid phase (water) is governed by the following terms:

$$\frac{dC}{dt} = KL \frac{A}{V} (Cs - C)$$

where:

– C: concentration of DO (g/m^3);

– KL: liquid film coefficient for oxygen (ms^{-1});

– A: surface area associated with volume V, over which transfer occurs (m^2/m^3);

– Cs: saturation concentration (g·m^{-3});

– t: reaction time (s).

Saturation concentration CS (mg·L^{-1}) is given by

$$CS = 14.652 - (0.41022)T + (0.00791)T^2 - (0.000077774)T^3$$

Temperature T is expressed in degrees Celsius.

The term A/V is the surface area per unit volume. The value of KL is given by the following equation, valid for air bubble diameters smaller than 0.25 mm:

$$KL = 0.28 \, D^{2/3} \left(\frac{\mu}{\rho} \right)^{-1/3} \sqrt[3]{g}$$

where:

- D (m²/s): O_2 diffusion coefficient, that is, $1.16793 \times 10^{-27} \, T_K^{7.392}$;

- μ and ρ: dynamic viscosity and density of water;

- g: gravitational constant;

- T: temperature in degrees Kelvin.

In a cascade, the parameters to be determined are the highest point (where the O_2 concentration is the lowest), the lowest point (where O_2 concentration is the highest), the number of transfer steps, each exchange step height, water sheet thickness, etc.

The O_2 concentration at the cascade outlet is given as $(mg \cdot L^{-1})$

$$Cf = C_{O2} = CS - \frac{CS - Co}{R}$$

R depends on the cascade's total height, inlet flow, height of the water outlet channel, spillway load and temperature.

Aeration efficiency is given by

$$E = \frac{C_{O2} - Co}{CS - Co} = 1 - (1 - k)^n$$

where:

- $Cf = C_{O2}$: outlet DO concentration;

- Co: inlet DO concentration;

- CS: O_2 saturation concentration;

- n: number of steps;

- k: oxygenation percentage obtained after each step.

A value of E equal to 1 means that the transfer made it possible to oxygenate water until saturation.

For a maximum drop height Hd (<10 cm), the spillway length L is given as

$$L(m) = \frac{Qwater}{1.77 \times 3{,}600 \times Hd^{1,5}}$$

The power of aeration R for each cascade is given by

$$R = (100 - P_{up})/(100 - P_{down})$$

where:

– P_{up}: percentage of DO upstream of the cascade compared to the final goal's saturation value:

$$P_{up} = C_{upstream}/Cf \times 100$$

– P_{down}: percentage of DO downstream of the cascade compared to the final goal's saturation value:

$$P_{down} = C_{downstream}/Cf \times 100$$

Cf: final concentration of DO.

Every elementary cascade with a cascade height Hc produces a unit rate R (approximately) equal to

$$R = 1 + a \times b \times Hc/2$$

where:

– a: 1.00 (moderately polluted water) to 1.25 (heavily polluted water);

– b: 1.00 (free fall) or 1.30 (stepped spillway);

– Hc: unitary cascade height.

REMARK.– For safety reasons, we will consider b = 1.00. By integrating (1) and (2), for moderately polluted water, we obtain

$$(100 - P_{up})/(100 - P_{down}) = 1 + (Hc/2)$$

where:

$$P_{down} = 100 - [(100 - P_{up})/R]$$

3.2.1. Data

The data are as follows:

– Flow of water to be reoxygenated Q: total flow: 340 $m^3 \cdot h^{-1}$

– Initial concentration of DO (outlet Biodagen for biological denitrification) Cin 0 $mg \cdot L^{-1}$ O_2 (between 0 and 1)

– Available cascade height: 2 m

– Final O_2 conc. Cf: 5.4 $mg \cdot L^{-1}$ O_2

– Drop height (overflow blade) Hd: 8 cm (lower than 10 cm)

– Spillway length = 340/(1.77 × 3,600 × Hd1.5): 2.36 m

– Slope = 45°

– Unitary cascade height Hc = 0.30 m (greater than 0.20 m)

– Step width = 0.30 m

– Initial P_{up} = 0%

– Hct is the total height of the cascade with Hct = Hc × n (N: number of steps)

The calculation parameters are given in Table 3.1.

3.2.2. Goals

– Downstream oxygen conc. (in relation to saturation) P_{down} > 50%.

– Unit rate R of an elementary cascade = 1 + 1 × 1 × 0.30/2 = 1.15.

3.2.3. Results

Nc	P_{up} (%)	P_{down} (%)	C_{down} (mg·L^{-1})	Hct (m)
1	0	13	1.4	0.3
2	13	24.3	2.6	0.6
3	24.3	34.2	3.7	0.9
4	34.3	42.8	4.6	1.2
5	42.8	50.3	5.4	1.5
6	50.3	56.7	6.1	1.8

Table 3.1. *Cascade parameter calculations*

Therefore:

– minimum number of cascades = 5;

– total cascade height = 1.50 m.

The design parameters are taken from the on-site results (Table 3.2).

Parameters	Formulae	Example
Temperature (°C)	T	15°C
Flow to be treated ($m^3 \cdot h^{-1}$)	Q	500
O_2 concentration in the water to be treated ($mg \cdot L^{-1}$)	Co	2
Oxygen saturation concentration ($mg \cdot L^{-1}$) at T°C	CS	10.08
Spillway load ($m^3 \cdot h^{-1} \cdot m$ spillway)	$q = 50–100\ m^3 \cdot h^{-1} \cdot m$ (manufacturer data)	$75\ m^3 \cdot h^{-1} \cdot m$
Spillway length (m)	L = Q/q	6.7
Number of steps	n = 3–12	4
Unitary step height (m)	Hm = 0.3 – 0.4 (manufacturer data)	0.4
Total step height (m)	Htm = Hm × n	1.6
Unitary step width (m)	l = 0.4–0.6 (manufacturer data)	0.5
Water sheet height (m)	$h = \left(\frac{0.5644\ q}{3600}\right)^{0.666}$	0.050
Unitary cascade height (m)	Hc = Hm + h	0.45
Total cascade height (m)	Htc = HC × n	1.8
Water height discharge channel (m)	He	0.1
Cascade surface (m²)	S = L.l ($0.020–0.05\ m^2/m^3 \cdot h^{-1}$)	13.4
O_2 deficiency coefficient at 20°C	R20: EXP ($0.0785.Hc1.31.q0.428.He0.310$)	1.089
O_2 deficiency coefficient at T°C	RT: EXP (LnR20 (1+0.0168 (T-20))	1.081
O_2 concentration at cascade outlet ($mg \cdot L^{-1}$):	$Cf = C_{O2} = CS – ((CS-Co)/RT)$	4.68
Aeration rate (%)	$Tx = C_{O2}/CS$	46.4%
Average increase per step	Aeration rate/n	11.6%

Table 3.2. *Cascade aerator design example*

A quick estimate can be obtained using the design parameters below.

The total height of the cascade can be quickly estimated by the following ratio:

$$H = \frac{R - 1}{0.198\,(1 + 0.046T)}$$

$$R = \frac{CS - Co}{CS - Cf}$$

where:

– CS: oxygen saturation concentration at temperature T °C mg·L^{-1};

– Co: O_2 concentration in water before aeration, mg·L^{-1};

– Cf: O_2 concentration in water after aeration, mg·L^{-1}.

For example, at 15 °C, the oxygen saturation concentration CS is 10.08 mg·L^{-1}.

For an inlet O_2 concentration of 2 mg·L^{-1}, and a required outlet concentration of 5 mg·L^{-1}

$$R = \frac{10.08 - 2}{10.08 - 5} = 1.59$$

The cascade height is given as

$$H = \frac{1.59 - 1}{0.198\,(1 + 0.046 \times 15)} = 1.76 \text{ m}$$

3.3. Aeration by fine bubble diffusers

Air insufflation can be an interesting solution when ozone or oxygen is not available. Aeration columns can be packed or work with atmospheric pressure and must be equipped with fine bubble diffusers (Figure 3.3). Because the kinetics are more advantageous than that of ozone, the contact time will not exceed 1–2 min. Compared to cascades, this solution achieves a higher dissolution rate without significant pressure loss. However, achieving the same dissolution rate involves a higher cost (both in terms of investment and operation). For this reason, it will only be adopted due to a technical need or at the customer's express request. Compared to membrane diffusers, porous diffusers ensure a higher transfer efficiency and a longer lifespan. Disc or dome models are better suited for small columns than tubular ones.

3.3.1. *Air diffusers*

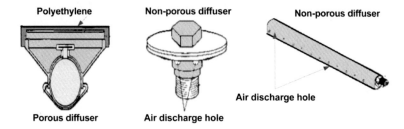

Figure 3.3. *Examples of porous and non-porous diffusers. For a color version of this figure, see www.iste.co.uk/gaid/watertreatment1.zip*

3.3.1.1. *Calculation of the quantity of air needed*

The oxygen transfer rate (OTR) is given by

$$rO2 = \frac{dm}{dt} = KT \, (CS - C)$$

K_T is the oxygen transfer coefficient. It depends on the temperature according to the following ratio

$$K'T = KT_{20} \, (1{,}204)^{T-20}$$

The quantity of air is obtained by

$$Qair = 0.00353 \, \frac{Qwater \, (CS) 20°C}{E(1{,}024)^{T-20}} \left(Ln \, \frac{CS - Cf}{CS - Co} \right)$$

The air flow Qair is expressed in $m^3 \cdot s^{-1}$, the water flow to be treated in $m^3 \cdot s^{-1}$, whereas E conveys the transfer efficiency, depending on the aeration system used. E may vary between 0.70 and 0.95. CS is the oxygen saturation concentration $(mg \cdot L^{-1})$ at temperature T °C, Co is the oxygen concentration in water before aeration, $mg \cdot L^{-1}$, whereas Cf is the oxygen concentration in water after aeration, $mg \cdot L^{-1}$.

3.3.2. *Air insufflation by oxytube*

3.3.2.1. *Data*

– Water flow to be re-oxygenated Q: total flow

– O_2 initial concentration Co = 0 mg·L^{-1} O_2

– O_2 final concentration Cf = 5.4 mg·L^{-1} O_2

3.3.2.2. Diffuser calculation

– Type of fine bubble diffusers: oxytubes

– Standard oxygen to be supplied = $340 \times 5.4/1000 = 1.84$ kg O_2·h^{-1}

– Water height H = 4.0 m

– Oxytube density d = 2 linear m/basin m^2

– Unitary air flow: 8 Nm^3·h^{-1}·m^{-1} oxytubes

– Transfer efficiency: 22 (for d = 1) + 4 (for d = 2) = 26%

– Air flow to be supplied: Qair: $1.84/(0.285 \times 0.26) = 24.8$ Nm^3·h^{-1}

– Total length of diffusers: Ldif: $24.8/8 = 3.1$ m

– Nominal length of an oxytube = 0.50 m

– Number of oxytubes: $3.1/0.5 = 6$

3.3.2.3. Aeration basin

– Basin surface area: $3.1/2 = 1.5$ m^2

– Minimum volume: $1.5 \times 4 = 6$ m^3

– Retention time: $6 \times 60/340$ = approximately 1 min

3.4. Stripping

The equilibrium of a species x between its gaseous phase and its liquid phase can be represented by

$$x(g) \leftrightarrow x(L)$$

The equilibrium between the gaseous form of x and the dissolved (liquid) form is expressed by Henry's law, with the coefficient carrying the same name Hx

$$px = Hx\ [x(L)]$$

where:

– px: partial pressure of x in the gaseous phase (atm);

– [x(l)]: concentration in the liquid phase (mol·L^{-1} or M).

Henry's constant Hx is therefore expressed in $mol \cdot L^{-1} atm$ or $mol \cdot kg^{-1} \cdot atm^{-1}$. The more soluble the gas, the stronger Henry's constant is. Henry's law only applies to diluted solutions, which is consistent with the concentrations generally observed in water. The higher the Henry's constant, the easier the stripping of dissolved gas or solvent.

Henry's law is valid for a single solute dissolved in a single solvent. If the solvent is not pure but contains other components (especially other solutes), Henry's constant changes and depends on the mixture's composition. Nonetheless, Henry's constant associated with each molecule indicates its facility to be degassed, even if several molecules are present at the same time in the water. Therefore, to remove supersaturations in dissolved gas (CO_2), the presence of gases such as hydrogen sulphide (H_2S) or methane (CH_4), volatile compounds such as chloroform, trichloroethylene, etc., it is necessary to promote the equilibrium of dissolved gases with the air by increasing the air–liquid contact surface as much as possible. The most common methods are those previously described as cascades and shallow diffuser aeration. There are also packed degassing towers.

3.4.1. *Stripping tower design*

The mass balance results in (Figure 3.4)

Accumulation = inlet load – outlet load + degassing

$$\frac{\partial C}{\partial t} \Delta V = QCh + \Delta h - QCh + r\Delta V$$

where:

– $\partial C / \partial t$: concentration variation as a function of time $g \cdot m^{-3} \cdot s^{-1}$;

– ΔV: volume (m^3) differential element;

– h: height (m) differential element;

– Q: water flow, $m^3 \cdot s^{-1}$;

– r: mass-transfer rate per unit volume and time $g \cdot m^{-3} \cdot s^{-1}$;

– C: concentration of the volatile compound, $g \cdot m^{-3}$.

This equation becomes

$$\frac{\partial C}{\partial t} = \frac{Q}{V} \frac{\partial C}{\partial h} + r$$

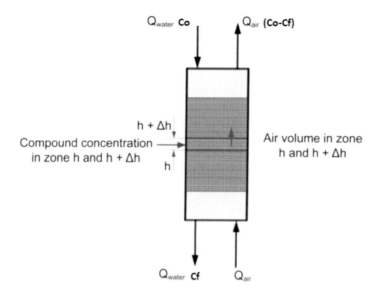

Figure 3.4. *Balance of concentrations and mass transfer in the stripping tower. Co: inlet concentration of the volatile compound in the water, $g \cdot m^{-3}$; Cf: outlet concentration of the volatile compound in the water, $g \cdot m^{-3}$; Cf, air: outlet concentration of the volatile compound in the air, $g \cdot m^{-3}$; Cf, air = Co – Cf, $g \cdot m^{-3}$ For a color version of this figure, see www.iste.co.uk/gaid/watertreatment1.zip*

The transfer velocity is

$$r = -K_{La} (C_{water} - C_{air})$$

where:

– K_{La}: volumetric mass-transfer coefficient depending on T °C and water characteristics, ls^{-1};

– C_{water} and C_{air}: concentrations of the volatile compound in water and in air, gm^{-3}.

At equilibrium, we obtain

$$\frac{\partial C}{\partial h} = \frac{KLa}{Q}(Cwater - Cair)$$

By integrating h between 0 and the support height H, we have

$$\int_0^H dh = \int_{Cf}^{Co} \frac{dCwater}{Cwater - Cair}$$

Since Cair depends on Henry's constant (KH)

$$Cair = \frac{Qwater\ P}{Qair\ KH}(Co - Cf)$$

where:

– Cf: outlet concentration of the volatile compound in water after stripping, $g \cdot m^{-3}$;

– P: pressure in atm (= 1 atm);

– Qair: air flow, $Nm^3 \cdot s^{-1}$;

– Qwater: water flow to be treated, $m^3 \cdot s^{-1}$.

By determining f

$$f = \frac{(Co - Cf)}{Cair} = \frac{Qair\ KH}{Qwater\ P}$$

The support media height is then given as

$$H = \frac{Qwater}{S\ KLa}\left(\frac{f}{f-1}\right) Ln\left[\frac{\left(\frac{Co}{Cf}\right)(f-1)+1}{f}\right]$$

where S is the stripping tower surface, m^2.

The transfer height is called Hz

$$Hz\ (m) = \frac{Qwater}{S\ KLa}$$

where S is the exchange surface (m^2).

That is,

$$H = Hz\left(\frac{f}{f-1}\right) Ln\left[\frac{\left(\frac{Co}{Cf}\right)(f-1)+1}{f}\right]$$

The number of transfer units (NTU) is given as

$$\text{NTU} = \frac{f}{f-1} \, Ln \, \frac{\left(\frac{Co}{Cf}\right)(f-1)+1}{f}$$

The total column height (Ht) is given as

Ht = Hz × NTU.

The minimum air/water ratio depends on the type of compound to be stripped. This is given by

$$\frac{Qair}{Qwater} = \frac{(Co-Cf)}{KHCo}$$

Co and Cf are the stripping column inlet and outlet concentrations, respectively, whereas KH is Henry's constant.

For example, the stripping of trichlorethylene (72 µg·L^{-1}) at 10°C for a flow of 8,200 m^3·j^{-1} is calculated as follows:

– trichloroethylene: C_2HCl_3;

– efficiency needed: 90%, that is, an outlet concentration of 7.2 µg·L^{-1};

– volumetric mass-transfer coefficient KLa: 0.0128 s^{-1};

– Henry's constant: 0.116 atm;

– P = 1 atm.

$$\frac{Qair}{Qwater} = \frac{(72-7.2)}{0.116 \times 7.2} = 7.76$$

The minimum airflow is therefore Qair = 8,200 × 7.76 = 63,632 m^3·j^{-1}. A safety margin of 3 is taken in the event of a variation in temperature and compound concentration, that is, 3 Qair/Q = 23.3. That is

Qair = 3 × 7.76 × 8,200 = 190,896 m^3·j^{-1}

f = 0.116 × 23.3 = 2.70

For a flow rate of 80 mh^{-1} (i.e. 0.0166 ms^{-1}), the total tower surface area is given as

$$S = \frac{8,200}{24 \times 3,600 \times 0.0166} = 4.27 \text{ m}^2 \text{ with a 2.33 m diameter}$$

The transfer height is written as

$$Hu\ (m) = \frac{8,200}{24 \times 3,600 \times 4.27 \times 0.0128} = 1.73\ m$$

The number of transfer units

$$NUT = \frac{2.70}{2.70-1}\ Ln\ \frac{\left(\frac{72}{7.2}\right)(270-1)+1}{2.70} = 3.47\ m$$

The total height is given as

Ht = 1.73 × 3.47 = 6 m.

3.4.2. Description

Stripping (at room temperature) comes down to extracting gases dissolved in water to turn them into the gaseous phase. To do this, a stripping gas is used, which is the oxygen in the air. Cocurrent (air–water) or countercurrent (water–air) contact in a vertical tower packed with a material such as Biodagen or plastic materials such as rings (Rashig, Pall, Levapac, etc.).

A degassing (or degasser) tower is made up of a column packed with an inert material or plastic rings held together by a grid at the base, allowing as much air as possible to pass through the tower (Figure 3.5).

The basic rules are as follows:

– an effective height (eH) of material between 1 and 3 m;

– support material size measuring between 1 and 5 cm;

– diameter of the column tower (D) limiting the water flow rate to 100 m·h^{-1};

– suitable equipment for annual cleaning of the base material.

The air is introduced through the tower's bottom, circulates against the water current (ascending direction), which streams over the packing, and is then extracted at the column's top. In practice, degassing towers make it possible to obtain an efficiency greater than 90%, depending on the implementation conditions and the physical characteristics of the compounds to be removed. The air is distributed via a power feed, which distributes the flow uniformly. In countercurrent systems, the water to be treated is distributed above the packing by means of a distributor.

A perforated floor provides support for the packing and avoids any risk of material breaking loose. The compounds thus extracted from the liquid phase are entrained by a strong air current. Before being released into the atmosphere, the air passes through a demister to remove the water droplets and to limit the risk of plume formation at the stripping outlet.

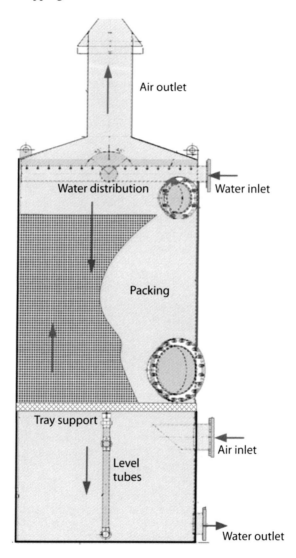

Figure 3.5. *OTV-Veolia countercurrent stripping tower. For a color version of this figure, see www.iste.co.uk/gaid/watertreatment1.zip*

The efficiency of degassing towers is affected by hydrodynamic variations since these variations are difficult to stabilize in the case of variations in the water flow to be treated. In addition, the gas/water flow ratio must be respected:

– if the gas flow becomes exceedingly large in relation to the water flow, the water flow is affected and greatly reduced. The material therefore becomes overwhelmed;

– if the water flow is too large in relation to the gas flow, the passage of the gas may be affected, thus resulting in back mixing within the gas stream.

3.4.3. CO_2 removal

Potabilization processes often include a free CO_2 removal stage when this is present in high concentrations. This stage can be set up at the beginning of the line to reduce the aggressiveness of water when evacuating the CO_2 (which could result in a pH increase) or to facilitate the preremineralization stage when adding alkaline reagents. This stage is also scheduled at the end of the process when the implementation conditions for the treatment plant require operating at acid pH (advanced coagulation-flocculation), high pressure membranes, etc.

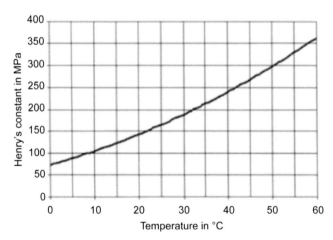

Figure 3.6. *Evolution of Henry's constant for CO_2 depending on temperature. For a color version of this figure, see www.iste.co.uk/gaid/watertreatment1.zip*

The efficiency of CO_2 stripping depends on the packing height and diameter, flow rate, and air/water ratio. For inlet CO_2 concentrations smaller than 30 mg·L^{-1}, an air/water ratio of 5 and a packing height of 1.2 m make it possible to obtain a CO_2 residual smaller than 5 mg·L^{-1}. For inlet CO_2 concentrations between 30 and 70 mg·L^{-1}, an air/water ratio of 10 and a packing height of 1.2–3 mm make it possible to obtain a CO_2 residual smaller than 5 mg·L^{-1}. For inlet CO_2 concentrations between 70 and 180 mg·L^{-1}, an air/water ratio of 15–20 and a packing height of 3–4.5 m make it possible to obtain a CO_2 residual ≤5 mg·L^{-1}. Henry's constant for carbon dioxide is 34 mol·kg^{-1}·atm^{-1} at 20°C, which justifies stripping this gas (Figure 3.6).

3.4.4. Tetrachlorethylene and trichlorethylene removal

These chlorinated solvents are easily removed by stripping due to their high Henry constant (trichloroethylene 550 mol·kg^{-1}·atm^{-1}) with air/water flow ratios greater than 8. Aeration, in contrast, is moderately effective. For 1,2-dichloroethane, the removal efficiency is limited (lowest Henry's constant). During stripping, chlorinated solvents are transferred into the air. It is advisable to take into account their possible environmental impact (see local legislation) and to consider possible granular activated carbon (GAC) adsorption of solvents in the gaseous phase.

3.5. Synthesis of aeration systems

Table 3.3 presents a summary of the various aeration systems commonly used in drinking water treatment systems.

Cascade	Tray aerator	Mass contact aerator	Air diffusion	Spray aerator
Description				
Creates convenient turbulence using a stepped cascade	Water flows thanks to gravity along a series of perforated trays	Packing tower with countercurrent air and water flow (Raschig rings, Pall rings, etc.)	Air diffusion via oxytubes in a tank	Spraying water on a material
Design criteria				
Spillway load: 50–100 m^3·h^{-1}·m spillway crest Step height: 30–40 cm Number of steps depends on treatment goals Approach velocity: >25 m/h	Distribution on the head tray: uniform Gap between trays > 15 cm Number of units: depends on goals	Water velocity: 50–100 m^3·m^{-2}·h^{-1} Air/water ratio: 3–10 N·m^3·m^{-3} Packing height: 1.0–3 m	Diffuser network in an aeration tank (fine bubbles) Contact time: 1–3 min	Fine bubble diffuser network

Table 3.3. *Characteristics and criteria of different aeration systems*

3.6. References

Baylar, A., Bagatur, T., Emiroglu, M.E. (2007). Aeration efficiency with nappe flow over stepped cascades. *Proceedings of the Institution of Civil Engineers – Water Management*, 160(1), 43–50.

Baylar, A., Kişi, O., Emiroğlu, M.E. (2008). Aeration efficiency estimation in stepped cascade aerators using neutral network approach. *Natural and Applied Sciences Civil Engineering*, 3(2), 361–371.

Chanson, H. (1994). Hydraulics of nappe flow regime above stepped chutes and spillways. *Australian Civil Engineering Transactions, IEAust*, CE36(1), 69–76.

Grace, G.R. and Piedrahita, R.H. (1993). Carbon dioxide control with a packed column aerator. In *Techniques for Modern Aquaculture*, Wang, J.K. (ed.). American Society of Agricultural Engineers, St Joseph.

Gulliver, J.S., Thene, J.R., Rindels, A.J. (1990). Indexing gas transfer in self-aerated flows. *Journal of Environmental Engineering, ASCE*, 116(3), 503–523.

Haarhoff, J. and Cleasby, J.L. (1990). Evaluation of air stripping for the removal of drinking water contaminants. *Water SA*, 16(1), 13–22.

Kavanaugh, M.C. and Trussell, R.R. (1980). Design of aeration towers to strip volatile contaminants from drinking water. *J. Am. Water Works Assoc.*, 72(12), 684–692.

Nakasone, H. (1987). Study of aeration at weirs and cascades. *ASCE Journal of Environmental Engineering*, 113(1), 64–75.

Onda, K., Takeuchi, H., Okumoto, Y. (1968). Mass transfer coefficients between gas and liquid phases in packed columns. *J. Chem. Eng. Jpn.*, 1, 56.

Piedrahita, R.H. and Grace, G.R. (1989). Carbon dioxide removal in a packed column aerator. *ASAE Annual Meeting*, American Society of Agricultural Engineers, St Joseph.

Rathinakumar, V., Dhinakaran, G., Suribabu, C.R. (2014). Assessment of aeration capacity of stepped cascade system for selected geometry. *International Journal of Chem. Tech. Research*, 6(1), 254–262.

Tatewar, S.P. and Ingle, R.N. (1999). Nappe flow on inclined stepped spillways. *Journal of the Institution of Engineers*, 79, 175–179.

Tebbut, T.H.Y. (1972). Some studies on aeration in cascades. *Water Research*, 6, 412–425.

Toombes, L. and Chanson, H. (2000). Air–water flow and gas transfer at aeration cascades: A comparative study of smooth and stepped chutes. *Proceedings of the International Workshop on Hydraulics of Stepped Spillways*, Zurich.

Sherwood, T.K. and Holloway, F.A.L. (1940). Performance of packed towers: Liquid film data for several packings. *Trans. Am. Inst. Chem. Eng.*, 36, 39–69.

Worm, G.I.M., Mesman, G.A.M., van Schagen, K.M., Borger, K.J., Rietveld, L.C. (2009). Hydraulic modelling of drinking water treatment plant operations. *Drinking Water Engineering and Science*, 2, 15–20.

Coagulation–flocculation

The most commonly used procedure for treating surface water includes coagulation, flocculation, settling, filtration and disinfection. This combination of processes is used for treating a wide range of raw water qualities. The first stages of this chain (coagulation, flocculation and settling) are all the more important because they influence the effectiveness of the following treatment stages, and disinfection in particular.

Not only does coagulation–flocculation make it possible to remove turbidity and suspended particles (including algae) but also to reduce color and organic matter and to remove microorganisms (bacteria, viruses, parasites, etc.). To properly understand the mechanisms of these two stages, it is useful to dwell on the concept of colloidal particles, which constitute an important parameter in the composition of raw water and require complete removal.

4.1. Colloidal matter

Due to their minuscule dimensions, colloids are subject to significant diffusion and sediment very slowly. For this reason, the role of interparticle forces and interactions is important for colloid stability. Thus, colloids are said to be stable if they are resistant to aggregation, whereas they are considered unstable if aggregation is quickly obtained.

In general, colloids fall into two categories: hydrophilic and hydrophobic particles. Examples of hydrophilic particles include proteins, starch, synthetic polymers, etc.

Hydrophobic particles, on the other hand, have little affinity with water: clay, rubber, latex, etc. Although these particles are unstable in the thermodynamic sense, they can be kinetically stable due to interparticle repulsions. In most cases, repulsive forces are electrical, and the general properties of colloidal particles are characteristic of a large developed surface and feature a distribution of negative charges across the entire surface.

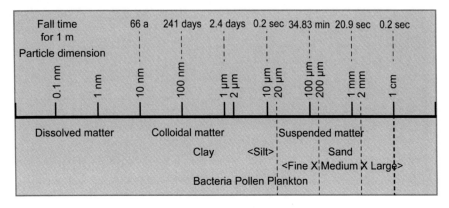

Figure 4.1. *Classification of particle types found in water*

Figure 4.1 illustrates the fact that the smaller the particles are, the longer the time required for them to settle. Therefore, the only way to settle them is to encourage their agglomeration to make up larger particles.

Particles suspended in water are subject to opposing forces that vary depending on the distance between such particles. The potential energy of interaction between two particles is the sum of the van der Waals attraction energy and the electrostatic repulsion energy related to colloid-surface charges. In general, at usual surface water pH values (pH between 5 and 8), the colloid surface is negatively charged. Thus, when two particles approach one another, several types of interactions occur, which can positively or negatively influence the coagulation–flocculation phenomenon. These interactions affect the collision frequency between particles, and flocculation effectiveness will essentially depend on the collision of the particles with one another to form an aggregate. It is therefore obvious that, if there is a strong repulsion between particles, the probability of forming an aggregate becomes very low, and even if flocculation slowly takes place, its overall efficiency remains low.

One of the first actions of the water engineer is to ensure a reduction in interparticle repulsions via the addition of chemical reagents, so that collisions between particles are as numerous as possible.

The two major types of particle interactions are van der Waals attraction forces and electrical repulsion forces. van der Waals forces result from the interaction between instantaneously induced dipoles in unpolarized, uncharged molecules. These dipoles originate from the fluctuations in the electron cloud of the molecule's atoms. They are both attractive and repulsive forces. They have weaker intensities than interatomic bonds and electrostatic (or ionic) forces and cover a short range. They nonetheless play an essential role in physical phenomena at the molecular level, such as capillarity, surface tension and particle "bonding" after collision.

Repelling electric forces (Figure 4.2) underlie the stability of colloids. It is convenient to neutralize them. The transition from a stable state of suspension to an unstable state is called destabilization, whose processes are known as coagulation and flocculation.

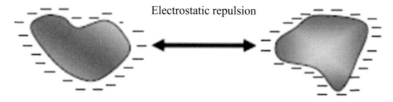

Electrostatic repulsion

Figure 4.2. *Electrostatic repulsion between two colloids. For a color version of this figure, see www.iste.co.uk/gaid/watertreatment1.zip*

However, this destabilization procedure induced by neutralizing the electric charges of the colloidal particles using a coagulant is called coagulation. On the other hand, the process whereby large flocs are agglomerated and clustered through the addition of high molecular weight polymers is called flocculation. In this way, at the very beginning of the process (and by means of various complex phenomena), coagulation makes it possible to flocculate colloidal particles, that is to say, to cause them to agglomerate. Afterward, flocculation promotes the agglomeration of particles, which then become decantable.

4.2. Coagulation

Coagulation is an essential stage in traditional water treatment because it enables the destabilization of colloidal particles. The forces of attraction between colloidal particles, known as van der Waals forces, are neutralized by the electrostatic repulsions of surface charges, which prevent particles from coming together and forming agglomerates. This is why the main mechanism to control the stability of hydrophilic and hydrophobic particles is electrostatic repulsion. The reason for this is that hydrophobic surfaces can include an excess of cations or anions producing an electric barrier that induces a repulsion of the particles having an identical surface potential. In addition to electrical repulsion, a suspension can be stable due to the presence of adsorbed water molecules, which provide a physical barrier by preventing particles from colliding. Both positively and negatively charged ions are attracted to the surface of each particle and can either be bound to the surface or distributed into the surrounding solution. In this manner, these two opposing forces (electrostatic attraction and ionic diffusion) produce a diffuse ionic cloud around the particles. The coexistence between an initially charged surface and the supply of counterions in relation to diffusely distributed co-ions in the solution is known as the electric double layer.

The mechanism can be summed up as follows:

– Helmholtz assumes that the solid's surface (particle) is covered with (generally negative) electric charges, which then form a continuous and homogeneous film. This creates an electrical imbalance that is compensated by the addition of positive charges coming from the solution. These charges are arranged opposite the ones present on the particle, forming a fixed double layer.

– The distance d between the two layers is approximately an order of magnitude of the molecular dimensions. Gouy and Chapman hypothesize a constant rearrangement between the two surfaces. Indeed, thermal agitation tends to disperse the atomic arrangement admitted by Helmholtz and to create a diffuse layer, with a predominance of positive charges. This charge density or excess is a continuous function of d, rapidly decreasing from the surface and asymptotically reaching zero at a sufficiently large distance. On the other hand, in Helmholtz's theory, the potential varies linearly with distance until it reaches zero.

– Stern associates the two hypotheses: Helmholtz's fixed layer and Gouy and Chapman's diffuse layer. His suggestion is that some of the ions with opposite charges form a fixed layer with the surface of the particle under the influence of electrostatic phenomena, whereas a second part of the ions, which are subjected to thermal agitation and Brownian motion (random motion of particles), forms a diffuse layer (Figures 4.3A).

– Stern observes a linear decrease inside a rigid layer and then a decrease tending toward an asymptote within the diffuse layer. He assimilates the first layer to Helmholtz's (with thickness delta) and the second layer to Gouy-Chapman's (with thickness d). He thus proposes the presence of two main layers: an internal layer (called Stern layer), in which the adsorbed ions are localized on the particles' surface, and an outer layer (called the Gouy–Chapman layer), made up of ions with opposite charges (counterions).

– The stability of colloidal suspensions is strongly influenced by the potential of the Stern layer (Figure 4.3B). Since it cannot be measured directly, the zeta (ζ) potential is used. It is the potential difference between the dispersion medium and the layer of stationary fluid attached to the dispersed particle.

– It is the only means for characterizing the properties of the double layer and is a key indicator of colloidal dispersion stability. A high zeta potential confers stability, meaning that colloid dispersion will resist aggregation. When the potential is low, attraction forces can overcome this repulsion and flocculation can take place. Thus, colloids with high zeta potential are electrically stabilized, whereas colloids with zero zeta potential tend to coagulate or flocculate

$$\zeta = 4\pi\, q\, d/D$$

where:

– ζ: zeta potential;

– q: amount of charge per unit area;

– d: layer thickness where a charge gradient is observed;

– D: dielectric constant of the liquid.

The goal of electrically destabilizing colloidal particles is to trigger the neutralization of surface charges, either by compressing the double layer (charge neutralization mechanism) or via the adsorption of a flocculant on the particles' surface (bridging mechanism).

The thickness δ of the ion cloud surrounding the particles is obtained because of Debye Huckel's formula (in nm at 25°C)

$$\delta = \frac{0.31}{\sqrt{\Sigma C Z^2}}$$

where:

– C: molar concentration of each species;

– Z: species valence.

When δ decreases, particles can get closer, in which case van der Waals forces are greater than electrostatic forces. This promotes particle aggregation.

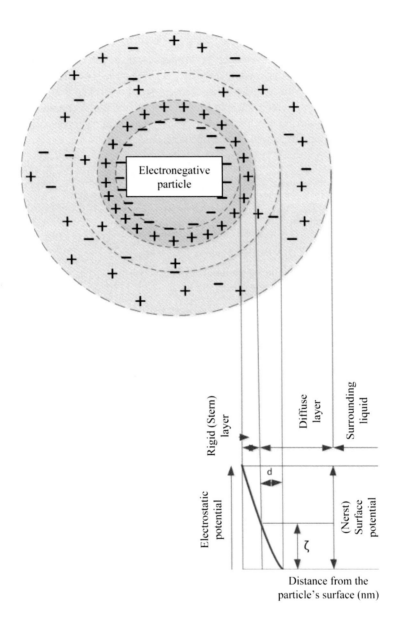

Figure 4.3A. *Diagram illustrating the charge distribution and Stern–Gouy–Chapman model. For a color version of this figure, see www.iste.co.uk/gaid/watertreatment1.zip*

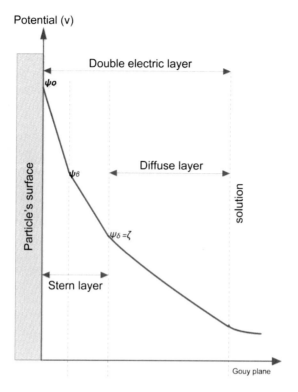

Figure 4.3B. *Diagram illustrating charge distribution and Stern-Gouy-Chapman model. For a color version of this figure, see www.iste.co.uk/gaid/watertreatment1.zip*

When added to water, coagulants (made up of hydrolyzable metal ions such as Al^{3+} and Fe^{3+}) quickly hydrolyze and form insoluble precipitates that destabilize negatively charged colloidal particles such as clay, algae, spores and viruses. This destabilization enables particle aggregation by means of chemical and van der Waals-like interactions.

Briefly stated, the traditionally identified elementary mechanisms to explain particle agglomeration are as follows:

– double-layer compression;

– charge adsorption and neutralization;

– formation of a colloid-trapping precipitate;

– creation of intraparticle bridging.

4.2.1. *Double-layer compression*

When the double layer is compressed, repulsive forces are reduced. Particles approach one another (Figure 4.4) and are attracted by means of van der Waals forces. The addition of ions such as Al^{3+} and Fe^{3+} releases high valence cationic species.

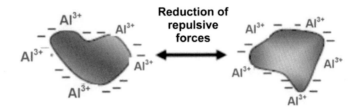

Figure 4.4. *Reduction of repulsive forces through the addition of high valence ions. For a color version of this figure, see www.iste.co.uk/gaid/watertreatment1.zip*

Taking the example of aluminum sulfate $(Al_2(SO_4)_3$, the dissolution reaction is as follows:

$$Al_2(SO_4)_3 + H_2O \;\rightarrow\; 2Al^{3+} + 3SO_4^{2-}$$

When the pH is low (<3.5), aluminum exists mainly in the Al^{3+} form, or rather as $Al(H_2O)_6^{3+}$ since Al^{3+} tends to hydrate by surrounding itself with six water molecules. At higher pH levels, hydrolysis reactions take place and form the following hydroxo-complexes (monomers). Depending on the pH, hydrolysis reactions occur successively. For example:

$$[Al(H_2O)_6]^{3+} + H_2O \quad\rightarrow\quad [Al(H_2O)_5OH]^{2+} + H_3O^+$$

$$[Al(H_2O)_5OH]^{2+} + H_2O \quad\rightarrow\quad [Al(H_2O)_4(OH)_2]^+ + H_3O^+$$

$$[Al(H_2O)_4(OH)_2]^+ + H_2O \quad\rightarrow\quad [Al(H_2O)_3(OH)_3] + H_3O^+$$

$$[Al(H_2O)_3(OH)_3] + H_2O \quad\rightarrow\quad [Al(H_2O)_2(OH)_4]^- + H_3O^+$$

In succession, these reactions form hydroxyl complexes leading aluminum to precipitate in the form of a relatively stable floc. In equilibrium with solid $Al(OH)_3$, the dissolved species (Figure 4.5) are essentially Al^{3+}, $Al(OH)^{2+}$, $Al(OH)_2^-$ and $Al(OH)_4^-$. The system's isoelectric point is approximately pH 6.

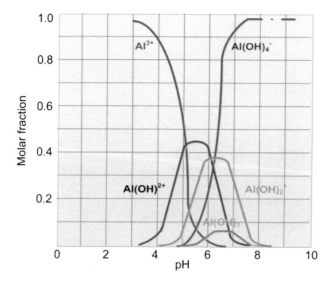

Figure 4.5. *Different forms of Al complexes depending on pH. For a color version of this figure, see www.iste.co.uk/gaid/watertreatment1.zip*

The activity of these ions induces a double-layer compression (and in particular, of the diffuse layer) toward the particle. Because of this, van der Waals attraction forces predominate, therefore enabling colloids to approach one another.

An increasingly concentrated salt solution decreases its zeta potential in absolute values, at the time its diffuse layer shrinks. Electroneutrality is reached at a lower distance from the colloid. From this distance, van der Waals attraction forces become strong enough to agglomerate two meeting colloidal particles.

4.2.2. Adsorption and interparticle bridging (flocculation)

Figure 4.6 shows the order in which these phenomena occur. Tardat-Henry shows that floc formation can be explained in different ways, depending on environmental conditions.

This double-layer compression neutralization phenomenon is involved in floc formation by means of cationic polymers and due to the ion adsorption/ neutralization phenomenon. The second neutralization stage involving coagulant adsorption (multivalent ions) on the colloid decreases the colloid's zeta potential.

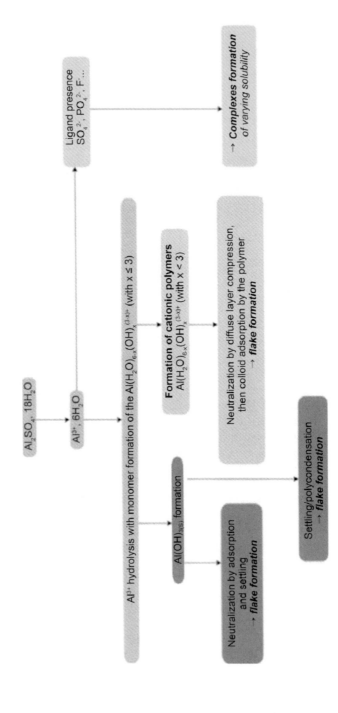

Figure 4.6. *Aluminum's transformation throughout the various coagulation/flocculation processes. For a color version of this figure, see www.iste.co.uk/gaid/watertreatment1.zip*

According to the Schultz–Hardy rule, the efficiency of monovalent, divalent and trivalent ions falls within a ratio of 1/100/1000. For drinking water applications, the effectiveness of Na^+, Ca^{2+} and Al^{3+} is 1/60/700, respectively. This means that Al^{3+} is expected to be 700 times more effective than Na^+ for charge neutralization. This is the reason why Al^{3+} and Fe^{3+} ions are most often used as coagulants.

It should also be noted that a trivalent coagulant (Al^{3+}, Fe^{3+}) is 10 times more effective than a divalent ion (Fe^{2+}) for decreasing the zeta potential. In other words, following the Schulze–Hardy theory, 10 times as many divalent ions should be dissolved to obtain the same result. Na^+ and Ca^{2+} salts are used as pH adjusting reagents, which stimulate the coagulation process. When adding the coagulant with 3+ ions, the zeta potential is zero (or even negative) because 3+ ions have compensated much of the colloid's negative charge. Thus, the electrostatic forces radiate over a shorter distance. Colloids can now approach one another to the point where van der Waals attraction forces prevail. In addition to the ionic forms of Al^{3+} bound to the colloid's surface, its hydrated forms are also involved in the coagulation–flocculation process. At acidic pH values, when the monomer $Al(H_2O)_5OH^+$ concentration is high enough, it tends to polymerize into larger molecules by establishing OH double bridges between the two Al atoms (Figure 4.7).

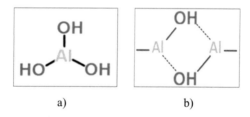

a)   b)

Figure 4.7. *(a) Al(OH)₃ configuration and (b) formation of OH double bridges between two atoms from (a). For a color version of this figure, see www.iste.co.uk/gaid/watertreatment1.zip*

The formation of polymeric species thus includes $Al_2(OH)_{24}^+$, $Al_3(OH)_{45}^+$, $Al_4(OH)_{162}^+$, $Al_6(OH)_{153}^+$ and $Al_{13}(OH)_{327}^+$ until $Al_{54}(OH)_{14418}^+$. These highly charged species can play an important role during coagulation. First, they neutralize colloids by significantly increasing the ionic strength of the medium. Second, they adsorb several colloids on their surface depending on the availability of reactive functional groups. Particles serve as precipitation seeds around which hydroxides agglomerate. These particles are thus trapped by the insoluble precipitates of metal hydroxides.

Coagulation is thus representative of all mechanisms for destabilizing a colloidal dispersion.

4.3. Flocculation

Flocculation is defined as the mechanism through which colloid flocs are caused to clump together and involves several phenomena. After destabilizing the colloidal particles by adding a coagulant – either through double-layer compression or by adsorption – flocculation promotes the agglomeration of destabilized particles into microflocs and then into decantable flakes. Flocculation is effective if the particles are able to agglomerate into increasingly large flocs. Two phenomena make it possible to understand and model the probability of flocs meeting up, depending on whether these are present as small flocs (beginning of flocculation) or as well-formed flocs (at the end of flocculation). The main path leading to floc formation is the conformation of a hydroxide precipitate ($Al(OH)_3$ or $Fe(OH)_3$). In addition, this agglomeration is facilitated by a particle transport phenomenon: thermal agitation (perikinetic flocculation) and the mechanical agitation of water (orthokinetic flocculation). The flocculation mechanism enables particles to communicate sufficient kinetic energy to cross the energy barrier. The flocculation mechanism itself involves perikinetic flocculation in the case of Brownian motion and orthokinetic flocculation in the case of laminar motion. Physicochemical flocculation makes it possible to lower the energy barrier by reducing electrostatic repulsive forces.

4.3.1. Perikinetic flocculation

The flocculation of small particles (smaller than 0.1 μm in diameter) is obtained through diffusion. The flocculation rate depends on the particles' rate of diffusion. The main particle aggregation mechanism is Brownian motion. This aggregation procedure is called microscale flocculation or perikinetic flocculation. After a period of a few seconds, microflocs vary in size, changing their diameter from 1 μm to approximately 100 μm.

It is necessary to imagine a multitude of colloid and/or microfloc particles. The meeting probabilities between these small particles are solely due to the thermal agitation of water molecules, following a random displacement (Brownian movements) and can be modeled emulating Brownian motion. Smoluchowski's idea was to imagine that there is one stable central particle, from which it is possible to calculate the number of particles that will meet with it as a function of time.

Smoluchowski's expression is given as:

$$Ln\frac{N}{No} = -\frac{4}{3}\alpha Vp\, Gt$$

where:

- N and No: number of free colloidal particles at time t and to;

- α: effective collision frequency factor;

- Vp: particle volume per volume of suspension;

- G: velocity gradient;

- t: contact time.

If parameter α is equal to 1, an interparticle shock produces agglomeration and therefore flocculation. The Vp element is constant – unless there is an external intervention – because there is neither creation nor disappearance of matter. The velocity gradient G is only an average value of the particles' specific velocities in the solution. Flocculation presents good quality if the Ln N/No ratio is small. In this case, N is smaller than No. There are therefore fewer free particles at time t than at time to. Any increase in parameters leads to a decrease in this ratio. The strategy to obtain good flocculation can be summed up as an increase in the factors: contact time t, particle volume Vp and velocity gradient G. The following equation shows the evolution in the number of floc particles for particle sizes smaller than 10 μm (colloids, microflocs):

$$\frac{dN}{dt} = -\frac{4}{3\,\mu}\,\alpha\,k\,T\,N^2$$

where:

- k: Boltzmann constant (1.38×10^{-23} J·K^{-1});

- T: temperature (K);

- N: particle concentration at time t (number of particles·m^{-3});

- α: fraction of effective shocks (giving rise to the agglomeration of two particles). It is the ratio between effective collisions and total collisions;

- μ: dynamic viscosity of water (Pa·s).

By integrating the expression of the number of particles at time t, we have

$N = 1/(1+(t/t_{1/2}))$ and $t_{1/2} = 3\ \mu/8kT\alpha No$

where $t_{1/2}$ corresponds to the time required for the concentration of particles to be reduced by half.

Note that the higher the number of initial particles, the more effective the shocks and the more $t_{1/2}$ is reduced. In the case of low turbidity, this explains the implementation of sewage sludge recirculation to increase the likelihood of encounters.

4.3.2. *Orthokinetic flocculation*

The main flocculation mechanism for particles greater than 1 μm diameter is based on the flocs mixing and aggregation conditions. This mechanism is known as macroscopic or orthokinetic flocculation. Mechanical mixing is required to achieve orthokinetic flocculation. It causes uneven shear forces on the floc, and part of the floc may be broken. After a mixing time, equilibrium in floc size is obtained, reaching a balance between the formation and rupture mechanisms.

Floc particles with different sizes settle at different velocities. These differences cause particles to collide and flocculate.

The addition of chemical flocculants is common practice. The addition of a polymer is performed after adding metal salts. This enables the formation of a floc that is then "bridged" by the polymer.

Perikinetic flocculation has shown that the lower the particle concentration, the longer it will take for particles to flocculate. For example, this time may be 3–4 h if the particle concentration is in the range of 10^9 per unit volume. When flocs reach the critical size of 10 μm, thermal agitation only plays an extremely minor role in particle displacement. Only the mechanical agitation of water corresponding to a weak turbulent regime can "cause" the particles to move and to make flocs grow.

The law that governs floc agglomeration is the following, and it shows that the law has the same "form" as the one governing perikinetic flocculation:

$$\frac{dN}{dt} = -\frac{2}{3}aN^2d^3G$$

where:

– α: fraction of effective shocks (giving rise to the agglomeration of two particles). It is the ratio between effective collisions and total collisions;

– N: particle concentration at time t (number of particles per m³);

– G: velocity gradient (Pa·s);

– d: particle diameter (m).

After integration with No and N for to and t, respectively, we obtain

$$Ln\frac{N}{No} = -\frac{2}{3}\alpha NoGtd^3$$

Therefore, the result is as follows:

$$N = 1/(1+(t/ t_{1/2})) \text{ and } t_{1/2} = 1/\alpha Nod^3G$$

Here, $t_{1/2}$ corresponds to the time required for the concentration of particles to be reduced by half. No and $t = 0$ correspond to the particle concentration and the time when perikinetic flocculation ceases and orthokinetic flocculation begins, respectively. It is useful to increase No to increase encounter probabilities. This is done in ballasted floc settlers (such as Actiflo, for example), where the concentration is increased by injecting microsand. The optimal point is obtained when the perikinetic flocculation rate is equal to the orthokinetic flocculation rate. We can deduce

$$G = \frac{\mu d^3}{(2\,kT)}$$

This means that when the particle diameter is small, the velocity gradient G must be high. This explains why the value of G during coagulation is high: it is due to the presence of microflocs. On the other hand, during the flocculation stage, G is lower, and we are in the presence of larger flocs (d is larger). In flocculation, the order of magnitude of G is $50–100\ s^{-1}$.

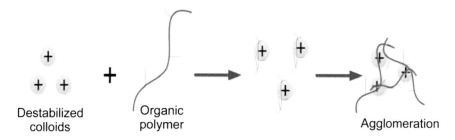

Figure 4.8. *Agglomeration principle with an organic polymer. For a color version of this figure, see www.iste.co.uk/gaid/watertreatment1.zip*

The pH of water should be in a range where salt solubility is minimal. The floc formed from $Al(OH)_3(s)$ – which has a positive Al^{3+} charge – will attract negatively charged colloids. After that, a colloid adsorption phenomenon follows on the Al^{3+} flocs, which will rapidly grow due to the mechanical trapping of colloids. On the other hand, $Al(OH)_3$ can also be adsorbed on colloids, neutralized, and then precipitated to form flocs. During the flocculation phase, no modification of the electric charges is performed on the flocs' surface. The addition of a high molecular weight polymer is described as a bridging mechanism among the microflocs formed during the coagulation stage (Figure 4.8). This favors an increase in the floc diameter and accelerates collision frequency. The flocculant's optimal concentration is more dependent on ionic strength than on molecular weight. However, flocculation efficiency can be affected by the dosing and mixing conditions in the flocculation tank. A low polymer dosage (0.1 mg·L^{-1}) is sufficient to flocculate high floc concentrations. This is provided that the mixing conditions are optimal because mixing too vigorously would tend to break the "bridges" created by the polymer. Too high a dosage ($>>1$ mg·L^{-1}) could lead to excessive coverage of flocs by the polymer and create a compact mass that is difficult to separate. This is why a mixer rotational velocity of 50 rpm is recommended. Thus, flocculation is representative of the transport mechanisms of destabilized particles leading to collision and aggregation and later enabling settling and separation. Non-ionic and anionic polymer chains with high molecular weight (10^4–10^7 g·mol^{-1}) adsorb on the free sites of several particles (Figure 4.9), bonding them to one another and thus forming decantable flocs.

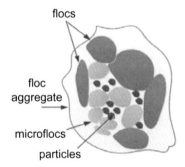

Figure 4.9. *Schematic representation of the different structuring scales inside a floc. For a color version of this figure, see www.iste.co.uk/gaid/watertreatment1.zip*

The structure of flocs formed during the destabilization of raw water colloids largely determines the efficiency of the clarification operation and the physical characteristics of the sediment obtained after settling (Figure 4.10).

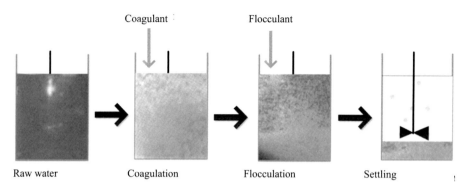

Figure 4.10. *Coagulation–flocculation sequential diagram. For a color version of this figure, see www.iste.co.uk/gaid/watertreatment1.zip*

4.3.3. *The influence of agitation*

The strategy to obtain appropriate flocculation implies either an increase in the contact time factors t or an increase in the volume of particles Vp and/or the velocity gradient G.

Any injection of additional particles of the microsand type (as is the case in Actiflo$^®$) or by sewage sludge recirculation in the flocculation area (Multiflo$^®$ Trio) increases the value of G. Moreover, for low turbidities, which would imply infrequent and random collision frequencies, the flocculation rate is increased by recycling the decanted floc in the flocculation tank (on average, 10% of the inflow) to improve the rate of floc formation. Then, denser and more voluminous flocs are obtained, which increases their sedimentation rate. Sewage sludge recirculation is particularly beneficial in low turbid and cold-colored waters (Figure 4.11).

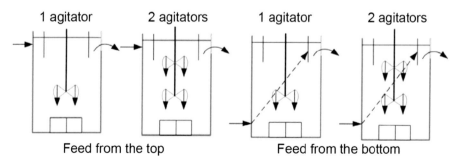

Figure 4.11. *Implementation of agitators in coagulation or flocculation tanks*

By approximately assimilating the number of particles to the water suspended solids (SSs) to be treated as well as by estimating the number of residual SS after flocculation time, we can arrive at an expression such as

$$Gt = -\frac{3}{4\,A}Ln\frac{N}{No}$$

where

$$A = 10^{-11}(SS)^2 + 8 \times 10^{-8}\,(SS) + 310^{-5}$$

A is dependent on collision efficiency and particle volume. For a value of G between 60 and 70 s^{-1}, we deduce the contact time t (s).

Coagulation requires a contact time in the range of 1–2 min for temperatures above 8°C and may require contact times of at least 4 min for temperatures below 2°C. The contact time for the flocculation stage varies between 10 min for a temperature above 20°C and rises to 30 min for a temperature below 5°C. This is because bridging kinetics are slowed down in cold water. In fact, the volume of the flocculation tank is always calculated for the lowest yearly temperature of the water to be treated. Except for Actiflo®, where the number of external particles contributed by microsand increases the value of No and helps in obtaining swift flocculation kinetics due to the highly frequent collisions between particles. In this case, the contact time is between 4 and 6 min, regardless of the temperature.

Camp proposed the product of Gt as a dimensionless number, which is used for evaluating flocculation adequacy. Camp numbers, Gt, ranging from 12,000 to 100,000 are considered to provide satisfactory flocculation. Experience shows it is recommended to be kept below 110,000. In fact, for excessively high G values, the floc formed undergoes mechanical shear, leading to its destruction.

G is obtained using the ratio

$$G = \sqrt{\frac{P}{V.\mu}}$$

where:

- G: mean velocity gradient (s^{-1});

- P: the actual dissipated power (W);

- μ: dynamic viscosity ($kg\cdot m^{-1}\cdot s^{-1}$);

- V: volume occupied by the fluid (m^3).

This definition of G is applicable to any type of hydraulic regime and is particularly dependent on temperature. For coagulation, it is necessary to spread all the coagulant quickly before the appearance of precipitates and to perform brisk agitation in a short period of time.

4.3.4. *G and t*

Gt is the product of the velocity gradient and time. The selection of G and t values for coagulation depends on the mixture's conditions and the coagulants. Since coagulation takes place by adsorption of the soluble hydrolyzed species on colloidal particles and trapping of the hydroxides thus formed, preliminary trials in jar tests provide a good indication of the best conditions to be implemented.

4.3.4.1. *The case of static mixers*

Mixers have two advantages (Figure 4.12):

1) there are no moving parts;

2) no external power source is needed.

On the other hand, their drawback is that the mixing intensity and the mixing time depend on the flow rate. They are often applied only for low coagulant doses and for temperature conditions that do not affect the coagulation process.

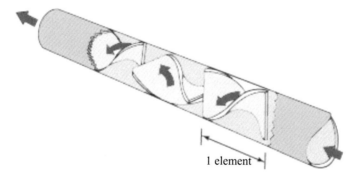

1 element

Figure 4.12. *Static mixer. For a color version of this figure, see www.iste.co.uk/gaid/watertreatment1.zip*

Gt is between 350 and 1700 for a contact time of 2–5 s and a maximum load loss of 0.6–0.9 m.

The dissipated power is given by:

$$P = \frac{\rho Q H}{\eta}$$

where:

- P: power in kW;

- ρ: specific mass of water: 9.807 kN·m^{-3} at 5°C;

- Q: flow in m^3·s^{-1};

- H: total load loss in m;

- η: efficiency (0.98–1.0).

4.3.4.2. Baffled flocculation tank

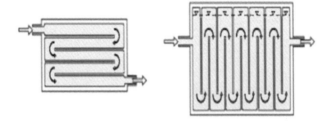

Figure 4.13. *Baffled flocculation tank. For a color version of this figure, see www.iste.co.uk/gaid/watertreatment1.zip*

Figure 4.13 shows the flocculation tank including horizontal baffles. The hydraulic flocculator is the baffled tank including a sinuous channel which is equipped with either around-the-end or over-and-under baffles. Each compartment plays the role of a small flocculator. The flocculation energy is derived primarily from the 180° change in direction of flow at each baffle. Generally, the minimum water depth is 1 m and the head loss across the flocculator is in the range 500 mm to 1 m. The contact time is 15–45 min, depending on the temperature, the coagulant et the dosage of the flocculant. The baffled basin has no mechanical or moving equipment and produces near plug flow. Short-circuiting is observed if the design is incorrect. Most head loss occurs at the 180° bends and therefore the value may be too high at the bends and inadequate in the straight channel. G value varies as the flow varies and can provide settlement of SS in the channel when the flow is too low. The recommended velocities to minimize settlement are 0.40 m·s^{-1} for high turbidity (>100 NTU), 0.30 m·s^{-1} for moderate turbidity (<60 NTU) and 0.25 m·s^{-1} for low turbidity (<10 NTU).

The number of baffles is given by:

$$n = \left[\frac{2\mu t}{\rho(1.44 + f)} \left(\frac{HLG)}{Q} \right)^2 \right]^{1/3}$$

where:

- n: number of baffles;
- H: depth of the water in the tank (m);
- L: length of the tank (m);
- G: gradient velocity (s^{-1});
- Q: flow rate ($m^3 \cdot s^{-1}$);
- t: contact time in the flocculator (s);
- μ: dynamic viscosity of the water ($kg.m^{-1} \cdot s^{-1}$);
- ρ: density of the water ($kg \cdot m^{-3}$);
- f: coefficient of friction of the baffles.

4.3.4.3. *The case of mechanical mixers*

There are several types of mechanical device for flocculation, common designs being the paddle type stirrers, mounted either horizontally or vertically in the flocculating tank, or axial flow impeller type stirrers which are used for high energy flocculation.

The dissipated power is given by

$$P = Np \; n^3 \; Dm^5 \; \rho$$

where:

- P: power in W;
- Np: power number;
- n: rotational velocity in $trs \cdot s^{-1}$;
- Dm: diameter of the mixer-stirrer in m;
- ρ: density of water.

The power required to obtain a velocity gradient G of 50 s^{-1} in a tank of 1000 m^3, treating a flow of 2,000 m$^3 \cdot$h^{-1} for a flocculation time of 30 min, is obtained by:

– temperature 5°C and 15°C;

– μ dynamic viscosity at 15°C:1.139 × 10^{-3} Ns·m^{-2} and at 5°C:1.518 × 10^{-3} Ns·m^{-2}.

We have:

$$P = G^2 \mu V$$

That is, at 15°C, P = 50² × 1.139 × 10^{-3} × 1000 = 28,475 W = 28.5 kW (or) 14.25 Wh/m^3 of treated water at 5°C, P = 50² × 1.518 ×10^{-3} × 1000 = 97,152 W = 37.95 kW (or) 18.9 Wh/m^3 of treated water:

– the Kolmogorov formula provides an estimate of floc length as a function of viscosity and gradient G:

$$IK = \left(\frac{\eta^2}{G^2}\right)^{1/4}$$

For the example above, we obtain

$$IK = \left(\frac{(1{,}139 \times 10^{-6})^2}{50^2}\right)^{1/4} = 1.49 \times 10^{-4}\ m = 149\ \mu m$$

– η kinematic viscosity at 15°C:1,139 × 10^{-6} m$^2 \cdot$s^{-1}.

4.3.4.4. *Calculation of the velocity gradient (G)*

4.3.4.4.1. General case

Table 4.1 summarizes the values usually implemented for velocity gradient G and contact time t. The agitation power is also indicated for each of the two stages.

	G (s-1)	t(s)*	Gt	Power W·m^{-3}
Coagulation	400–1,000	120–240	48,000–240,000	80–100
Flocculation	20–60	600–1,800	12,000–108,000	< 40 Wm^{-3}

*Variable flocculation time, depending on T (°C) and quality of the water to be treated.

Table 4.1. *Gt values are usually implemented for coagulation and flocculation*

4.3.4.4.2. Specific cases

	Coagulation			Flocculation		
	G (s^{-1})	t (min)	Gt	G (s^{-1})	t (min)	Gt
Low turbidity, colored waters	400–600	2	48,000–72,000	20–70	30	36,000–130,000
		4	96,000–144,000			
High turbidity (> 200 NTU)	400–1,000	2	48,000–120,000	30–80	20	36,000–96,000
Turbidity < 10 NTU or color, with sludge recirculation	400–600	2	48,000–72,000	70	30	126,000
		4	96,000–144,000			
Decarbonation	400–1,000	120–240	48,000–240,000	80–200	20	96,000–240,000

Table 4.2. *Usually implemented Gt values for different types of water*

Table 4.2 presents the velocity gradient values and the contact time implemented for different types of water. The energy required for coagulation is in the range of 80–100 W·m^{-3} per reactor. In contrast, flocculants are injected into a reactor with a longer retention time, and due to their shear sensitivity, the energies are lower: 40–50 W·m^{-3}. Recycling is automatic in sludge blanket clarifiers.

4.4. Coagulants

4.4.1. *Metallic coagulants*

The most commonly used metallic coagulants are divided into two categories: those based on aluminum and those based on iron. The popularity of two coagulants (aluminum sulfate and ferric chloride) is based not only on their effectiveness (in optimal operating conditions) but also on their attractive cost.

These two coagulants have some limitations, such as:

– pH adjustment before and after treatment, associated with preremineralization for freshwater;

– a sensitivity of ferric chloride for cold waters;

– higher sewage sludge production for ferric chloride;

– sometimes high dosage, depending on the chemical composition of the water to be treated.

When brought into the solution, aluminum sulfate and ferric chloride dissociate very quickly to form hydrated reaction products. Thus, the metal ion forms compounds such as $\{Al(H_2O)_6\}^{3+}$ and $\{Fe(H_2O)_6\}^{3+}$.

4.4.1.1. Aluminum-based coagulants

Aluminum salts are commonly used in water treatment. The main products marketed are aluminum sulfate, aluminum chloride, sodium aluminate and polymerized salts: aluminum polychloride and polychlorosulfate. The flocs formed by aluminum hydroxide are less dense and settle with more difficulty than the flocs formed by iron salts but have the advantage of being much less colorful. Commercial aluminum salts are generally characterized by their aluminum content, expressed in % Al_2O_3 (representative of the active "matter" contained) and by the product's "basicity", expressed by the molar ratio $(OH^-)/3(Al^{3+})$, which corresponds to the fraction of sites occupied by hydroxyl functions in relation to the total number of available sites. The base is the preneutralization degree of hydrogen ions during hydrolysis. It is calculated using the following equation, where [OH]/[M] is the molar ratio of the hydroxide bound to the metal ion, and ZM is the charge of the metal species

$$\text{Basicity (\%)} = \frac{100 \times [OH]}{[M] \times ZM}$$

Basicity affects the ratio between the polynuclear and mononuclear species present in the solution and provides an indication of the coagulant's alkalinity consumption.

4.4.1.1.1. Aluminum sulfate [Al$_2$(S0$_4$)$_3$, nH$_2$0]; n = 14–18

It is probably the most widely used and one of the most economical. Aluminum sulfate is produced by the reaction of caustic soda with bauxite or another aluminum-enriched mineral (such as kaolin) and then acidified with sulfuric acid. Its Al_2O_3 content generally oscillates between 8.5% and 17%. It is corrosive and available in solid or liquid form. Aluminum sulfate dissociates and reacts with water and forms a panoply of soluble mononuclear and polynuclear species (comprising one or more aluminum ions), as well as an amorphous aluminum hydroxide precipitate.

When adding an aluminum salt to water, the basic reaction is the precipitation of aluminum hydroxide and the release of acid

$$Al_2(SO_4)_3 \longrightarrow 3SO_4^{2-} + 2Al^{3+}$$

$$2Al^{3+} + 6\,H_2O \longrightarrow 2Al(OH)_3 + 6H^+$$

The solubility product of $Al(OH)_3$ is $[Al^{3+}]\,[OH^-]^3 = 10^{-33}$. The precipitation of $Al(OH)_3$ leads to the trapping of particles in suspension in water, whose dimension is greater than or equal to 1 μm. The mode of action of aluminum salts is based on the fact that the ion Al^{3+} is hydrated by six molecules of water and forms an octahedral ligand. Its effectiveness depends on the degree of aluminum hydrolysis. The hydrolysis reaction begins immediately after dosing the aluminum sulfate into water. Similar to any chemical reaction, the degree of hydrolysis depends on the raw materials, reaction time, reaction temperature, etc. Aqueous solutions of aluminum sulfate react with the alkaline carbonates present in the water and cause the formation of CO_2, a shift in the calco-carbonic equilibrium.

The theoretical effective hydrolysis pH range for aluminum sulfate is 6.0–6.5, with optimum pH values near 6.3. At optimum pH, solubility is minimal, and the maximum amount of coagulant is turned into solid floc particles. As the pH decreases below 6.0, positively charged and dissolved aluminum species are formed. When the pH is above 6.5, the concentrations of negatively charged aluminum species increase. The residual aluminum concentration in the produced water should be monitored.

4.4.1.1.2. Aluminum chloride [$AlCl_3,6H_2O$]

Hydrolysis reactions between water and metal salts produce hydrogen ions that lower the pH and consume alkalinity at a 1:1 equivalent. During the manufacturing process, neutralization of hydrogen ions with a basic solution leads to the formation of prehydrolyzed coagulants, such as those made from aluminum chloride. Aluminum chloride is usually presented as Al_2O_3 at 20%. Its pH and density are 2.5 and 1,300 kg·m^{-3}, respectively. It is generally available in the liquid form. This product is highly corrosive. Its use is optimal in the pH range of 5.5–7.8. Aluminum chloride is a good product for general coagulation uses. It is sometimes used in combination with organic polymers.

4.4.1.1.3. Sodium aluminate [$NaAlO_2$]

Sodium aluminate is an alkaline salt prepared by the action of aluminum oxides with sodium. It precipitates well in the pH range between 5 and 7. Sodium aluminate contains approximately 13% Al and can be combined with aluminum sulfate for specific difficult water conditions, such as highly colored waters and also for low-alkalinity waters. The joint use of these coagulants is called "double coagulation": it prompts faster precipitation, producing a denser floc. In addition, it enhances the control of pH and of water alkalinity. However, prior testing is recommended because in some cases coagulation is not as effective as expected. The hydrolysis reaction is as follows:

$$NaAlO_2 + 2H_2O \longrightarrow Al(OH)_3 + NaOH$$

4.4.1.1.4. Aluminum chlorohydrate

Aluminum chlorohydrate (ACH) generally refers to coagulants based on aluminum chloride with basicity values of approximately 83%. The minimum solubility pH of ACH (pH 6.0–7.2) is significantly higher than that of alum, which enables it to be effective at higher pH values without increasing aluminum residuals dissolved in water. It is more efficient at low temperatures and produces less sewage sludge.

4.4.1.1.5. Polyaluminum chlorides and inorganic aluminum polymers, etc.

A distinction is made between polyaluminum chlorides (PACl or PAC), polyaluminum chlorosulfates (PACS), polyaluminum sulfates (PAS) and polyaluminum silicate sulfates (PASS). Before their implementation, these polymers are obtained by making OH^- ions react on aluminum salts, which trigger a polymerization effect due to an olation reaction. These often lead to a better decantable floc. They both act by means of electrostatic discharge and colloid bridging. They constitute an improvement of the aluminum salts in that they promote a combined and immediate action of the coagulation process thanks to the numerous available Al^{3+} sites. They also favor a bridging action by virtue of their polymeric structure. TAC and pH are less affected by this type of coagulant. Despite being more expensive than non-polymerized aluminum salts, they have the advantage of requiring lower dosages than aluminum sulfate for an equivalent quality of water to be treated. They usually grant excellent quality in the treated water, a better cohesion of sewage sludge and low residual aluminum content. These products are particularly recommended for treating surface water (e.g. for removing turbidity) but are not very effective for removing color and organic matter. The most common commercial products are WAC, WAC HB, Aqualenc, PCBA (liquid) and

PACl (powder). WAC-type polyaluminum chlorides have a formula of the $Al_n(OH)_m Cl_{3n-m}$ type, with an m/n ratio between 1.35 and 1.80.

Polyaluminum chloride is written PACl to differentiate it from PAC (powder activated carbon). PACl is the product of the reaction between solid aluminum hydroxide $(Al(OH)_3)$ and 33% hydrochloric acid (HCl). In terms of solubility, PACl has a higher optimal pH value than alum, which enables it to form precipitates at higher pH values (pH 6.0–7.8). Due to its polymeric structure, PACl reacts faster than aluminum sulfate, which gives it a stronger coagulant power. The flocs formed are larger, which implies a faster sedimentation process. It consumes less alkalinity than many other coagulants. In most cases, there is no need to raise water alkalinity to achieve effective flocculation: no additional chemical reagents are needed to correct the pH.

These coagulants feature the following advantages:

– they are easy to handle and apply (liquid form);

– their efficiency is relatively independent from temperature;

– low acidity contribution in relation to the dose introduced, which translates as lime saving when compared to aluminum sulfate;

– their efficiency is relatively independent from overdosing;

In particular, they are used for the treatment of cold waters loaded with mineral colloids, poorly loaded with organic matter and slightly colored.

Polychlorosulfates of aluminum (PACS and PASS), with the formula $Al_n(OH)_m (SO_4)_3 Cl_{3n-m-2x}$, are offered with a 10–18% Al_2O_3 content. Depending on the case, they contain variable proportions of chlorides, hydroxides, sulfates and silicates, which give them stable functional properties. PACS presents high efficiency as a coagulant compared to conventional or simply prepolymerized coagulants. The introduction of silicate chains within the PACL structure increases its molecular weight due to the formation of aluminosilicate complexes. This improves particle aggregation and leads to larger flocs. They are being increasingly used because they are effective across a wide pH range (pH 6.0–8.0) and temperature and require lower dosages than aluminum sulfate. This is explained by the nature and the number of Al species present in this type of coagulant. They are preferred over Al sulfate for cold waters because they instantly form $Al(OH)_3$ flocs, which adsorb contaminants on the flocs' surface.

	Formulae	Al_2O_3 %	Densities at 20°C $kg \cdot m^{-3}$	pH at 20°C	Basicity %	Molecular masses $g \cdot mole^{-1}$
Aluminum sulfate	$Al_2(SO_4)_3$, nH_2O n = 14 liquid n = 18 solid	8.5% 17%	1,320 1,690	2.5 1.8	0.2% 0.3%	594 666
Aluminum chloride	$AlCl_3$	8.2%	1,250	–	–	133.5
Sodium aluminate	$Al(OH)_4Na$	18%	1,470	–	–	118
WAC Aluminum polychlorosulfate	$Al_n(OH)_mCl_{3n-m}$, n and m between 0.45–0.60	10%	1,200	3.2	>38.5%	Variable
HB-WAC Polyaluminum chloride sulfate	$Al_n(OH)_mCl_{3n-m-2k}(SO_4)_k$, n and m between 1.95–2.25	8.5%	1,170	<1	65–75%	Variable
AQUALENC F1 Al polychlorosulfate	$Al_2(SO_4)xCl_y(OH)z$	9%	1,220	2.6–3.0	62–72%	Variable
AQUALENC E Al basic chlorosulfate	$Al_2(SO_4)xCl_y(OH)z$	17%	1,380	0.4–0.8	37%	Variable
AQUALENC S Al basic chlorosulfate	$Al_2(SO_4)xCl_y(OH)z$	10%	1,220	2.0–3.0	70%	Variable
PAC 18	$Al_2(OH)nCl$ 6-n }x	18%	1,220	–	–	–

	Formulae	Al_2O_3 %	Densities at 20°C $kg \cdot m^{-3}$	pH at 20°C	Basicity %	Molecular masses $g \cdot mole^{-1}$
PAC 30	$Al_2(OH)nCl \; 6-n \}x$	30%	1,220	–	–	–
PAX XL3 Poly Al chlorohydrate	$Al(OH)x(Cl)y$	10%	1,240	2.5	70%	Variable
PAX XL10 Poly Al chlorohydrate	$Al(OH)x(Cl)y$	9.7%	1,220	1	70%	Variable
PAX XL 9 Al polyhydroxychlorosulfate	$Alw(OH)x(Cl)y(SO_4)z$	8.5%	1,220	2.5	–	Variable
PAX XL 60 Al polychlorosilicate	$Alw(OH)a(Cl)b(Si0x)c$	14.2%	1,300	1.5	70%	Variable
PAX X18	$Al(OH)x(Cl)y$	17%	1,360	1	43%	Variable
PASS polyhydroxysilicate Al sulfate	$Al(OH)3.24(Si0x)0.1(SO_4)1.58$	10%	1,340		54%	–
R PASS polyhydroxysilicate Al sulfate	$Al (OH)a(Si0x)c$	8.3%	1,270	3.2	50%	Variable
PACS Al polychlorosulfate	$Al_3(OH)4.95Cl3.55 (SO_4)0.25$	10%	1,190	–	50%	–

Table 4.3. *Physicochemical characteristics of Al-based coagulants*

Basic polyaluminum chloride (PCBA) differs from conventional aluminum sulfate by faster flocculation, proper elimination of organic matter, a lower treatment rate, approximately 45% less sewage sludge volume, a reduced risk of overdosing and the addition of generally unnecessary flocculation aids. However, its reduced stability preferably reserves it for large treatment facilities (in situ production) or for regional use.

4.4.1.2. Iron-based coagulants

The main iron salts used in coagulation are ferric chloride, ferric sulfate, ferric chlorosulfate and ferrous sulfate. Unlike aluminum salts, there are no polymerized salts with a high degree of basicity in the case of iron. Ferric salts, which are more loaded, have a better coagulating power than ferrous salts. The mode of action of iron salts, including ferric chloride, is well understood: in an acid solution, the ion Fe^{3+} is hydrated by six water molecules and forms an octahedral ligand. Depending on the pH, hydrolysis reactions occur successively:

$$Fe(H_2O)_6^{3+} + H_2O \rightarrow [Fe(H_2O)_5OH]^{2+} + H_3O^+$$

$$[Fe(H_2O)_5OH]^{2+} + H_2O \rightarrow [Fe(H_2O)_4(OH)_2]^+ + H_3O^+$$

$$[Fe(H_2O)_4(OH)_2]^+ + H_2O \rightarrow Fe(H_2O)_3(OH)_3 + H_3O^+$$

$$Fe(H_2O)_3(OH)_3 + H_2O \rightarrow [Fe(H_2O)_2(OH)_4]^- + H_3O^+$$

On the other hand, ferric hydroxide ions $[Fe(H_2O)_5OH]^{2+}$ tend to polymerize (Figure 4.15) with one another and form "oxo" bridges when present in sufficient concentrations.

In succession, these reactions form iron hydroxyl complexes and cause iron to precipitate in the form of a relatively stable floc. These reactions also lowered the pH due to the release of protons. Metal ions are bonded by hydroxide bridges, whereas the products of ferric ion hydrolysis neutralize the colloids' negative charges and therefore ensure their destabilization. In addition, the formation of polymerized compounds can yield large floc volumes that can then surround the colloidal particles and trap them.

The isoelectric point of the system is approximately pH 8: below 8, the positively charged form predominates while the negative species is dominant; when the pH is greater than 8, the use of a high dose of ferric salt often provides a rust color to the treated water: this is the main drawback of these products. However, this color can be easily eliminated by filtration in a basic medium. The optimum pH zone for this precipitation is between 4 and 8.5.

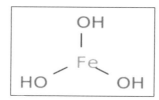

Figure 4.14. *Iron hydroxide structure*

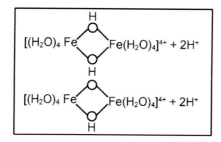

Figure 4.15. *Example of oxo bridges from iron-hydroxide ions*

4.4.1.2.1. Ferric chloride (FeCl₃, 6H₂O)

Ferric chloride is available in liquid or solid (crystalline or anhydrous) form. It is obtained by dissolving scrap iron in hydrochloric acid (formation of $FeCl_2$) and by oxidation with liquid chlorine. It is commonly used in liquid form, with an approximate 40–42% $FeCl_3$ content by weight. It also exists in the crystalline form, with a 60% $FeCl_3$ content by weight. It is highly corrosive. Ferric chloride aqueous solutions are highly acidic and react with the alkaline carbonates present in water. This reaction results in the formation of CO_2 and a shift in the calco-carbonic equilibrium. In addition to the coagulation of colloidal substances, ferric chloride enables the reduction of certain soluble organic compounds. However, the optimum pH for the reduction of organic matter is different and more acidic than the optimum pH required for turbidity removal. Fe^{3+} ions can be hydrolyzed and precipitated into $Fe(OH)_3$ at pH values above 4.5. The pH range extends between 5.5 and 8.5. However, establishing the optimum pH range and coagulant concentration ultimately depends on the characteristics of raw water and differs from one water source to another. The solubility product of $Fe(OH)_3$ is $[Fe^{3+}][OH^-]^3 = 10^{-38}$. Unlike Al hydroxides, ferric hydroxide does not dissolve once again in a basic medium. This makes it operational at basic pH for certain applications. These coagulants are used for removing color and organic matter under pH conditions below pH 5. Since they often contain high proportions of Mn^{2+}, it is necessary to acknowledge this concentration increase in flocculated water and to remove Mn^{2+} at the filtration stage by adding a small amount of sodium upstream. Currently, there are $FeCl_3$ coagulants that contain less than 50 mg Mn^{2+} per kg $FeCl_3$ compared to the classic ratio (600 mg Mn^{2+} per kg $FeCl_3$).

The low solubility of iron hydroxides (up to pH 8.5) often makes ferric chloride preferred to Al salt coagulants. However, they are less effective in cold water than Al-based coagulants. Finally, $Fe(OH)_3$ flocs are denser and larger than $Al(OH)_3$ flocs.

The $FeCl_3$ reactions with alkalinity or lime are as follows:

$$2\ FeCl_3 + 3\ Ca\ (HCO_3)_2 \rightarrow 2\ Fe(OH)_3 + 3\ CaCl_2 + 6\ CO_2$$

$$2\ FeCl_3 + 3Ca(OH)_2 \rightarrow 2\ Fe(OH)_3 + 3\ CaCl_2$$

One milligram of ferric chloride consumes 0.92 mg of alkalinity and generates 0.66 mg of sewage sludge.

Ferric chloride doses can range between 5 and 220 $mg \cdot L^{-1}$, depending on multiple factors, including concentrations of natural organic matter (NOM) and raw water quality.

4.4.1.2.2. Ferric sulfate ($Fe_2(SO_4)_3$, nH_2O)

Ferric sulfate is obtained by the reaction between ferric hydroxide and sulfuric acid. The hydration number is usually 2 or 3, although some forms can have up to nine molecules of water per molecule of ferric sulfate. It is available in the solid form (with 72–75% $kg/Fe_2(NA_4)_3$ kg and 20–21% Fe^{3+}) or the liquid form (with 40–42% $kg/Fe_2(SO_4)_3$ kg and 11.5% Fe^{3+}). However, this type of liquid coagulant can be offered with a significantly lower purity (30%). As ferric chloride, it is also corrosive. Like most ferric coagulants, it can be used within a pH range between 4 and 8.5 for applications in drinking water. Its mode of action is similar to that of ferric chloride:

$$Fe_2(SO_4)_3 + 9H_2O \rightarrow 2\ Fe\ (OH)_3 + 6H^+ + 3\ SO_4^{2-} + 3H_2O$$

The reactions of ferric sulfate with bases are as follows:

$$Fe_2(SO_4)_3 + 3Ca\ (HCO_3)_2 \rightarrow Fe(OH)_3 + 3CaSO_4 + 6\ CO_2$$

$$Fe_2(SO_4)_3 + 6NaHCO_3 \rightarrow Fe(OH)_3 + 3Na_2SO_4 + 6CO_2$$

$$Fe_2(SO_4)_3 + 3Na_2CO_3 + 3H_2O \rightarrow 2Fe(OH)_3 + 6CO_2$$

$$Fe_2(SO_4)_3 + 6NaOH \rightarrow 2Fe(OH)_3 + 3Na_2SO_4$$

$$Fe_2(SO_4)_3 + 3Ca(OH)_2 \rightarrow 2Fe(OH)_3 + 3CaSO_4$$

One milligram of ferric sulfate consumes 0.75 mg of alkalinity and generates 0.54 mg of sewage sludge in the form of iron hydroxide. Ferric sulfate behaves much like ferric chloride but is less used in water treatment plants. Typical dosages range from 10 to 250 mg·L^{-1}.

4.4.1.2.3. Ferrous sulfate (FeSO$_4$, 7H$_2$O)

Ferrous sulfate is the end product of the action of sulfuric acid on iron. This product is subject to a few constraints, including the mandatory maintenance of the solution at more than 10°C. It is available in solid or liquid form. It is most often offered in the form of crystals or granules containing approximately 20% Fe. Ferrous sulfate must be oxidized as Fe(OH)$_3$ to be usable for coagulation. It is optimally used at high pH values (pH > 7.7). The combination of ferrous sulfate and chlorine is performed upstream of the coagulation stage to produce ferric chloride and ferric sulfate, following the simplified reaction:

$$6 \text{ Fe (H}_2\text{O)}_7\text{SO}_4 + 3 \text{ Cl}_2 \rightarrow 2 \text{ Fe}_2(\text{SO}_4)_3 + 2 \text{ FeCl}_3 + 42 \text{ H}_2\text{O}$$

The molecules ferric chloride and ferric sulfate react in turn and provoke coagulation. The FeSO$_4$,7H$_2$O reactions with alkalinity or lime are as follows:

$$\text{FeSO}_4, 7\text{H}_2\text{O} + \text{Ca(OH)}_2 \rightarrow \text{Fe(OH)}_2 + \text{CaSO}_4 + 7\text{H}_2\text{O}$$

$$4\text{Fe(OH)}_2 + \text{O}_2 + 2\text{H}_2\text{O} \rightarrow 4\text{Fe (OH)}_3$$

The solubility product of iron hydroxide: Ks = $10^{-15.1}$.

4.4.1.2.4. Ferric chlorosulfate (FeClSO$_4$)

Ferric chlorosulfate is marketed in liquid form. This product is obtained by oxidizing ferrous sulfate with chlorine or through the reaction of acids on ferric oxides

$$6\text{FeSO}_4, 7\text{H}_2\text{O} + 3\text{Cl}_2 \rightarrow 6\text{FeClSO}_4, 7\text{H}_2\text{O}$$

The combination of ferrous sulfate and chlorine is performed upstream of the coagulation stage to produce ferric chloride and ferric sulfate.

Note that 1 mg·L^{-1} of FeSO$_4$ theoretically requires 0.13 mg·L^{-1} chlorine. In drinking water, this product is used to reduce turbidity.

	Formulae	Iron levels	Densities at 20°C kg·L^{-1}	pH at 20°C	Molecular masses g·mole^{-1}
FeCl$_3$ Ferric chloride	FeCl$_3$, 6H$_2$O	14–15%	1.44	1.8	162.5
FeSO$_4$ Ferrous sulfate	FeCl$_4$, 7H$_2$O	20.00%	0.9	1.5–2.2	152
Fe$_2$(SO$_4$)$_3$ Ferric sulfate	Fe$_2$(SO$_4$)$_3$, 9H$_2$O	12–25%	1.44–1.55	<2	400
PFS Polyferric sulfate	Fe$_2$(OH)$_{0.6}$, (SO$_4$)$_{2.7}$	12.2%	1.50	<2	382.4

Table 4.4. *Physicochemical characteristics of iron-based coagulants*

4.4.1.2.5. Al^{3+}/Fe^{3+} mixed products

The combined use of iron salts with aluminum salts can produce a synergistic effect for the treatment of difficult waters. An Al/Fe ratio = 0.3 associated with a pH between 5.8 and 6.3 can lead to a good quality of treated water, with a lower residual rate of metal hydroxides and requiring a lower total coagulant dose. Laboratory tests are necessary to verify this application. There are mixed mineral coagulants on the market that provide both Al^{3+} and Fe^{3+} ions. The purpose of using these mixed salts is to optimize the respective advantages of iron and aluminum salts. This is the case for mixed aluminum and iron sulfate, polyaluminum ferric chloride (PFAC), polyaluminum ferric sulfate (PAFS) and Prodefloc AFC, which is a mixture of polyaluminum chloride and ferric chloride. Their use must be verified according to local regulations.

	Formulae	Al$_2$O$_3$	Densities at 20°C kg·L^{-1}
Ferric and aluminum sulfate	(Al$_x$Fe(1-x))$_2$(SO$_4$)$_3$	26-40 gAl·kg^{-1} 6-21 gFe·kg^{-1}	Variable
Fe (III) and Al hydroxy chloride	AlFe$_x$(OH)aClb	5.3% Al	1.3

Table 4.5. *Characteristics of mixed Al/Fe coagulants*

4.4.2. Synthetic organic coagulants

Polymers used in water treatment generally have a low molecular weight (<500,000) and can be used as coagulants. They are often associated with metallic coagulants. The polymers used as coagulation aids are generally cationic compounds. They are frequently dosed in the range between 0.1 and 0.6 mg·L^{-1}. The most commonly found are one of two quaternary amines: polydiallyldimethylammonium chloride (polyDADMAC) or epichlorohydrin-dimethylamine (epi/DMA).

Organic coagulants are polyelectrolytes, that is, water-soluble polymers with high molecular weight and different ionicities, obtained when one or more monomers are polymerized. These polymers can be cationic, anionic, non-ionic or amphoteric. However, most organic coagulants are cationic, and their action is usually based on the specific chemistry of nitrogen. In fact, the additional pair of electrons available in nitrogen can form a fourth bond (quaternization) and induce a positive charge, which is capable of neutralizing the negative charges present on the surface of particles and colloids. Coagulants lacking non-quaternary amine groups, as well as other functional groups, are also used in water treatment.

Synthetic organic coagulants are used when colloids have a low zeta potential and water is composed of fine SSs. Indeed, the mode of action of these polymers by coagulation is limited due to their reduced number of positive charges. On the other hand, they are very effective for treating fine SSs by adsorption. In addition, they lead to a significant reduction in the volume of sewage sludge and therefore to an increase in its density. They are used as a partial replacement for mineral coagulants. pH regulation is not necessary since these coagulants only slightly modify pH, just like salinity. A synergy can be found with the joint use of a classic mineral coagulant and a synthetic organic coagulant. This combination makes it possible to produce a lower sludge volume (but a fairly sticky sludge, and therefore not suitable for all types of settling tanks) to reduce the amount of mineral coagulant required (40–60%) while increasing the effectiveness of coagulation on colloids and fine SS.

To make the most of the advantages each category of coagulants offers, suppliers use mixtures of organic coagulants and iron or aluminum salts to optimize various parameters, such as load availability, the doses needed, pH adjustment, the quality and the quantity of sludge produced.

Residual acrylamide monomers are present in polyacrylamide coagulants used in drinking water treatment. In general, the maximum authorized dose of synthetic acrylamide polymer oscillates between 0.6 and 1.2 mg·L^{-1} depending on the country, which greatly reduces their use as a main coagulant. This applies only to anionic and non-ionic polyacrylamides, but residual levels of cationic polyacrylamides may be higher.

4.4.2.1. Poly(DADMAC) and Epi/DMA

Poly(DADMAC) compounds are prepared by addition polymerization of acrylamide with a cationic monomer (diallyldimethylammonium chloride), and their average molecular weight is generally between 5×10^4 and 1.5×10^6. This cationic polymer is used as a coagulant for the reduction of turbidity and color, as well as for sewage sludge conditioning. Poly(DADMAC) is chlorine resistant, and its charge density is independent of pH since its charge is of the quaternary type. Its use is widespread, either alone or in combination with inorganic coagulants, including iron and aluminum salts.

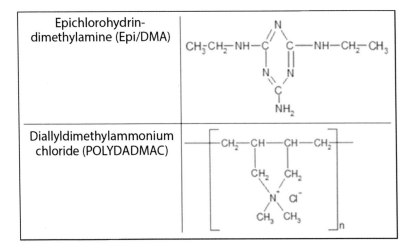

Figure 4.16. *Poly(DADMAC) and Epi/DMA formulae*

Epi/DMA polymers are formed by condensation polymerization between epichlorohydrin and dimethylamine, and their average molecular weight ranges from 10^4 to 10^5. As with poly(DADMAC), Epi/DMA is a quaternary polyamine used in a mixture with inorganic coagulants to optimize performance. It is often used for color reduction in drinking water treatment. These polymers act in a similar way to metal salts, that is, by destabilizing colloidal dispersions. The doses required are mainly between 0.5 and 10 $mg \cdot L^{-1}$, and an overdose may result in the system's restabilization and reduced coagulation efficiency.

4.4.2.2. Quaternary and tertiary polyamines

Several representatives of the large group of cationic quaternary ammonium polyelectrolytes, featuring the $-NR_4^+$ function, are also used as coagulants. Apart from polyamines with high cationic quaternary charge, organic coagulants having tertiary amine functional groups, such as $-NHR_3^+$, are also used for some specific applications.

4.4.2.3. Organic coagulants or mineral coagulants?

Organic coagulants form large and robust flocs very quickly. The concentration in sulfates or chlorides is not as significant as the one obtained with aluminum or iron salts. Dosage and sludge production are clearly lower. They do not require pH adjustment and their implementation is simpler. However, they are effective only against turbidity but much less effective against color and organic matter. All things considered, they are costly and are rarely used precisely due to these economic reasons.

4.5. Flocculants

4.5.1. Natural organic and synthetic flocculants

The most widely used organic flocculants are synthetic polymers with high molecular weights (up to 15 million). They have a linear structure, which makes them soluble in water. These polymers periodically carry ionized chemical groups, which therefore makes them carriers of electrical charges. For this reason, they are called polyelectrolytes. When they are dissociated, molecules become charged, either positively or negatively, after which they are referred to as cationic and anionic polyelectrolytes. When they are not ionizable, polymers are called non-ionic. Since these macromolecular chains are made up of carbon atoms joined by single bonds that behave as a succession of articulated segments, the flocculating power of polymers is related to the configuration of macromolecules within a solution. They are classified into two main categories: natural and synthetic polymers. Flocculation can be improved by producing more mechanically resistant flocs (better resistance to shear stress) and with a larger size, which also enhances settling. These polymers make it possible to capture already well-formed microflocs. The efficiency of these flocculants was assessed by means of jar tests. The main parameters for assessment are size, floc cohesion and settling rate.

4.5.1.1. Natural flocculants

Despite the fact that natural polyelectrolytes have the advantage of being free from toxic substances, the use of synthetic polyelectrolytes is more widespread. In general, they are more effective flocculants, mainly due to the control of their properties, such as the number and the type of charged units and molecular weight.

In addition, they are much cheaper than those made from natural sources. Natural organic flocculants are water-soluble polymers of animal or vegetable origin. Generally non-ionic, they can be chemically modified. Their molecular weight is lower than that of synthetic polymers, which gives them poorer flocculation properties. Their interest lies in their "natural", non-toxic, biodegradable character. The most widely used are starch, alginates and guar or xanthan gums. However, these products are only reserved for specific applications because they are expensive, rare compared to the size of the water treatment market (in the case of guar gums, for example) and have reduced effectiveness due to their short chain length. These polymers can be classified depending on their charge (anionic, non-ionic, cationic or amphoteric), size and ionic strength (0–100%).

Starch is a polysaccharide (polymer of glucose) of vegetal origin. The exact chemical nature of starch varies depending on its origin, but the most active forms of starch are those rich in monophosphoryled esters, as found in potatoes. Starch is a normally non-ionic macromolecule but can be modified or degraded to obtain anionic or cationic functions. It is a loosely bound, water-soluble polysaccharide. The anionic type is often obtained by carboxylic substitution. Starch (Figure 4.17) is not a homogeneous substance but is made of two substances: amylose and amylopectin.

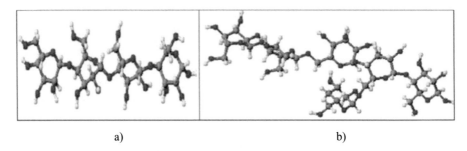

a)   b)

Figure 4.17. *Chemical formulae of amylose (a) and amylopectin (b), two constituents of starch. For a color version of this figure, see www.iste.co.uk/gaid/watertreatment1.zip*

The mechanism is identical to that of synthetic polymers. There are basically two modes of action for these products, which in certain cases can be combined:

– mechanism for neutralizing electric charges: flocculation takes place in the vicinity of a zero zeta potential and is all the lower as the molecular weight of the polyelectrolyte is low;

– adsorption and cross-linking (branching) mechanism: this is the exclusive flocculation mechanism for treating negative particles with anionic polymers. Macromolecular substances can be adsorbed on the surface of the dispersed solid phase in a limited number of sites. Flocculant molecules attached to the individual particles still have free active centers for adsorption. These centers attach themselves to sites on the surface of other particles. In this way, "bridges" are created between several particles, forming a three-dimensional network and giving rise to a flake.

Dosage with natural-origin polymers requires higher doses than with synthetic polymers. In fact, for flocculation to take place, it is necessary for polyelectrolyte molecules to be adsorbed on the surface of the particles that carry a charge of the same sign. Adsorption can be thwarted by electrostatic repulsive forces. In addition, for interparticle bridge formation to be possible, macromolecular chains (which may have one or more adsorbed segments) need to spread sufficiently within the liquid. This is achieved by the repulsions of the ionic groups located along the chain. It is therefore necessary to take into account two contradictory imperatives: to use sufficiently ionized macromolecules to have a suitable chain extension, and, on the other hand, the degree of ionization has to remain sufficiently low for adsorption to be possible. That being so, there is a given suspension, an optimum percent ionicity for the anionic polyelectrolyte to flocculate. Adsorption will thus be all the more thwarted when the particles' zeta potential is negative and when the polyelectrolyte's percent ionicity is high. This is why commercial products hardly exceed the value of 40% ionicity.

The interparticle "bridges" that are set up should be as long as possible. To achieve this, a product with the highest possible molecular weight is recommended. The theory predicts a linear relationship between the optimal dose of polyelectrolyte and colloidal matter concentration. Natural organic flocculants are water-soluble polymers with a lower molecular weight (from 10^6 to 10^7 Daltons) than synthetic polymers, which gives them poorer flocculation properties. It is useful to note that the spatial arrangement of a starch-based polymer is not linear as a polyacrylamide, since it is composed of branched chains (amylopectin) and linear chains (amylose). The ratio is 80% and 20% for Hydrex 3801 and 3843, respectively. This results in lower anionicity degrees for natural flocculants in the zeta potential range from –500 to –1500 $\mu eq \cdot g^{-1}$, instead of –1,000 to –4,500 $\mu eq \cdot g^{-1}$ for anionic polyacrylamides. For this reason, it is essential to properly match the overall characteristics of the natural flocculant with the characteristics of the raw water to be treated. Their interest lies in their "natural", non-toxic, biodegradable character.

The major advantage of natural polymers is that they easily accompany sustainable development concepts by virtue of their biodegradable nature (even when requiring higher dosages). The polymer composition in terms of the dissolved organic carbon to biodegradable dissolved organic carbon (DOC/BDOC) ratio does not influence the flocculation mechanism or the final quality of the treated water. One might think that the fact that a natural polymer is essentially made of a DOC/BDOC ratio very close to 1 could cause an increase in the final BDOC concentration. However, this is not the case for several reasons: the DOC present in the water to be treated is composed of humic substances and other types of molecules that are previously adsorbed on $X(OH)_3$ flocs before being fixed on polymers. The BDOC removal mechanism is identical to that of DOC since it is an integral part of it. Despite their good solubility in water, natural polymers leave no residue in water because their structure acts directly and globally on flocs by creating a solid mass that can be easily separated from water. Through this liquid/solid separation, all the BDOC in the natural polymer is entrained with the sewage sludge, without leaving any residue in the interstitial water. This is the reason why the BDOC analyzed in treated water is identical to that obtained with a synthetic polymer. In the same way that adding a synthetic organic polymer during flocculation does not increase the treated water's DOC, the addition of a natural polymer does not lead to an increase in BDOC because the mechanism for trapping and adsorbing flocs and particles engages their steric and electrical configuration, not their chemical composition.

Green polymers represent a very interesting alternative to synthetic polymers. Hydrex (Veolia) markets a range of modified starch-based products, such as Hydrex 3801 and 3843.

Gums are polysaccharides that have several functional groups, such as carboxyl groups. Guar gum is mainly obtained from the seeds of a leguminous plant belonging to the genus *Cyamopsis*. It has the ability to form hydrogen bonds with mineral surfaces. Guar gum is relatively insensitive to the pH and ionic strength of the effluent to be treated. Xanthan gum is a polysaccharide produced by a microorganism, *Xanthomonas campestris*.

Alginates are obtained from alginic acid, which is itself extracted from marine algae. They are mainly available in the form of sodium, potassium, ammonium and calcium salts. The essential constituents of their polymeric structure are mannuronic and glucuronic acid. When combined with ferric salts, they result in particularly effective flocculation products. They can also give good results with aluminum salts. A major drawback of most natural polymers lies in their rapid degradation in an aqueous medium, especially when the external temperature is above 20°C. The equipment (preparation tanks, in particular) must be cleaned regularly to avoid the risk of fermentation. Natural organic polymers are interesting because, in comparison with the use of synthetic organic polymers containing acrylamide monomers, no danger to human health due to their use has been identified.

4.5.1.2. Synthetic organic coagulants

The majority of synthetic flocculants are derivatives of acrylic monomers forming homopolymers or copolymers whose molecular weight can vary from a few hundred thousand to several million Daltons. These are water-soluble polyelectrolytes obtained by polymerization of one or more monomers and with a highly variable ionic charge, the main ones being polyacrylamides. However, in regard to their flocculant function, all these products always seem to act in accordance with charge neutralization mechanisms and interparticle bridging. The performances obtained with synthetic flocculants have made them occupy a largely predominant position compared to natural polymers.

Synthetic flocculants are polymers with a linear structure and high molecular mass (up to 15 million). When macromolecular chains carry ionized chemical groups (electric charges), they are called polyelectrolytes. They can be cationic (carrying positive charges) or anionic (carrying negative charges). The electrically charged groups of the same sign and distributed along the chain mutually repel one another and provide the macromolecule with the greatest extension in water. The parameters related to these macromolecules and taking part in flocculation are the chemical nature, the electric charges they contain, their molecular mass and their degree of ionicity.

Flocculants can be non-ionic, cationic, anionic or amphoteric. Adsorption phenomena or chemical interactions are involved, including with certain soluble organic molecules. In the case of negatively charged particles, the cationic polymer acts through its charge and neutralizes the particle's still available free sites. For this, it is immediately added after the coagulation stage. The action of cationic polymers can be explained by their strong adsorption on negatively charged particles and the consequent reduction of double-layer repulsion, thus enabling aggregation. The most effective cationic flocculants are those with high charge density.

The anionic polymer tends to act by interparticle bridging and is more effective when added after the formation of microflocs. As a complement for a mineral coagulant, an anionic polymer is generally used at a variable concentration between 0.05 and 0.6 $g \cdot m^{-3}$. On the other hand, for sewage sludge treatment, dosage is expressed in $kg \cdot ton^{-1}$ of dry matter and varies between 2 and 10 $kg \cdot t^{-1} \cdot DM^{-1}$. Flocculants are available as powders or emulsions. An excessive addition of flocculants causes system restabilization, thus making flocculation ineffective. The amount of flocculant added should be given special attention.

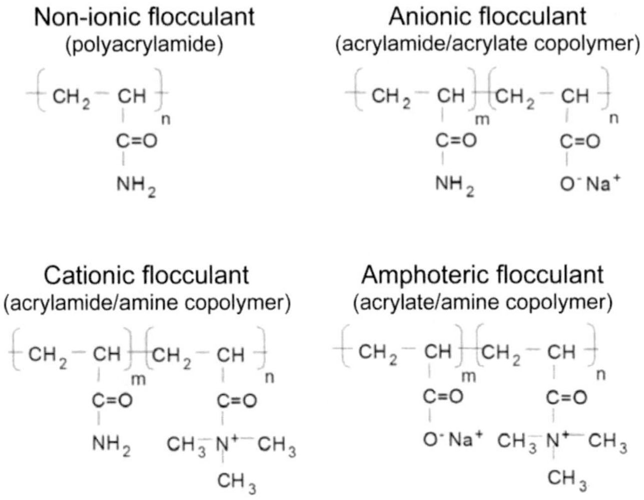

Non-ionic flocculant
(polyacrylamide)

$$\left[CH_2 - CH \right]_n$$
$$C=O$$
$$NH_2$$

Anionic flocculant
(acrylamide/acrylate copolymer)

$$\left[CH_2 - CH \right]_m \left[CH_2 - CH \right]_n$$
$$C=O \quad\quad C=O$$
$$NH_2 \quad\quad O^- Na^+$$

Cationic flocculant
(acrylamide/amine copolymer)

$$\left[CH_2 - CH \right]_m \left[CH_2 - CH \right]_n$$
$$C=O \quad\quad C=O$$
$$NH_2 \quad CH_3 \; N^+ \; CH_3$$
$$CH_3$$

Amphoteric flocculant
(acrylate/amine copolymer)

$$\left[CH_2 - CH \right]_m \left[CH_2 - CH \right]_n$$
$$C=O \quad\quad C=O$$
$$O^- Na^+ \quad CH_3 \; N^+ \; CH_3$$
$$CH_3$$

Figure 4.18. *Structure of organic polymers used as flocculants*

Polyacrylamide is often used with aluminum or iron salts and makes it possible to reinforce floc resistance to shearing. The recommended doses of polyacrylamide for conventional flocculation are usually $0.1-1.0 \ mg \cdot L^{-1}$. Anionic polyelectrolytes containing the carboxylic function are, by far, the most widely used type of polymer to function as anionic flocculants. Many of these polymers are derivatives of acrylamide and are prepared by polymerization and acrylamide hydrolysis. Epichlorohydrin (cationic polyelectrolyte) is synthesized by a polymerization reaction with ammonia or a primary or secondary amine.

4.5.1.3. *Natural or synthetic polymer?*

The main advantage of natural polymers is their availability as a renewable natural resource and the lack of evident toxicity or by-products in treated water in comparison with the acrylamide monomer obtained from polyacrylamides. Moreover, they are biodegradable. Unfortunately, they require dosages at least three times higher than those of synthetic polymers. Synthetic polymers perform well at low dosages, displaying chemical stability. On the other hand, their biological stability can also be a disadvantage because any residual in the treated water ends up at the consumer's tap. Finally, depending on the dosage used, monomer residues (acrylamide monomer type) can be found in the treated water. There is a strict standard for this parameter (less than $0.1 \ \mu g \cdot L^{-1}$).

4.5.2. Flocculation adjuvants

A flocculation adjuvant makes it possible to accelerate flocculation phenomena or to reinforce the flocs formed. In particular, they are used for treating water containing a low concentration of SSs.

Activated silica is an inorganic anionic polyelectrolyte prepared from sodium silicate after neutralization with sulfuric acid (Figure 4.19). The normal concentration use of activated silica is 5–10% of the mineral coagulant used, that is to say, 1–5 $mg·L^{-1}$ of SiO_2 and, more rarely, up to 10 $mg·L^{-1}$. Activated silica is used at the end of coagulation or at the start of flocculation for neutral or slightly acidic waters rich in organic matter. Although its use with an iron salt is ineffective (especially with an acidic pH), it works properly with aluminum salts, especially in cold water. One of the problems associated with this product is the propensity to form gels, which can block pumps and pipes. With this adjuvant, it is possible to obtain denser and more resistant flocs to accelerate sedimentation. The condensed structural formula is shown in Figure 4.19.

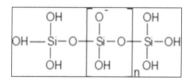

Figure 4.19. *Semideveloped formula of activated silica*

Bentonite-type clay (montmorillonite) is an aluminosilicate and negatively charged clay. Bentonite is often used with aluminum coagulants and offers satisfying results for doses of approximately 100 $mg·L^{-1}$.

4.6. Factors affecting coagulation and flocculation

The type of coagulant used influences the coagulation mechanisms that can be implemented. The type of coagulant is chosen depending on the characteristics of the water to be treated, in particular, its temperature and alkalinity.

4.6.1. Influence of water temperature

Temperature affects several important coagulation/flocculation parameters, including floc size, density, water viscosity and even pH, to a certain extent. Compared to metallic coagulants, it affects hydrolysis reaction rates for coagulation and bridging for flocculation. The reaction takes place rapidly in hot water (<10 min) but is slowed down in cold water (<8°C). To overcome this problem, the dosage and/or contact time between the water and the coagulation/flocculation reagents must be increased. For example, the reduction in turbidity by alum may be only 20%, or even less, when the water is at 1°C, compared to the same water at 25°C (>90%). In addition, flocs settle less quickly in cold water than in hot water.

The minimum and maximum solubility pH values of metal hydroxide precipitates are increased by a decrease in temperature, thus modifying the optimal operating conditions for each coagulation mechanism. On the other hand, the prehydrolyzed coagulants of metal salts are less sensitive to low temperatures, although they can still be affected up to 15%.

4.6.2. Influence of pH

pH is an important parameter for obtaining efficient coagulation. Each coagulant has an optimum pH range because, for each type of coagulant, the pH and the coagulant dosage are responsible for the mechanisms involved in particle coagulation. A pH range and coagulant dosage are associated with each type of coagulation mechanism. When coagulation occurs outside this optimal range, several disadvantages appear: increased dose needed, iron or aluminum release, reduced performance, etc.

In addition to its influence on metallic salts, pH has an influence on particle charge as well as on several ionizable substances. pH plays an outstanding role in coagulation because a balance must be reached between the pH needed for coagulation (colloid destabilization) and the pH required for flocculation (floc growth and particle trapping). It is sometimes necessary to increase or decrease the pH of water to improve coagulation and flocculation. Extreme pH values can affect polymers, for example, by breaking covalent bonds (in an acid medium) or by destroying the charge of the quaternary amine group (in a basic medium). These phenomena are irreversible. Depending on the system studied and the surrounding conditions, a variation in pH values can be positive or negative, since they have an impact on the coagulation and adsorption efficiency of organic compounds. For example, the coagulation of humic substances using ferric salts is highly dependent on the medium's pH. In general, the optimum pH for removing dissolved organic compounds is lower than that used for removing turbidity. However, it should be set in the interval where there is formation of a fairly solid floc for the adsorption phenomena to take place.

pH is affected by different types of coagulants to varying degrees. For example, metal salts have a significant impact on pH, and prehydrolyzed species have a lesser impact, whereas polymers practically do not change the pH of the water to be treated. pH adjustment with an acid or a base makes it possible to improve the performance of coagulation–flocculation. The chemicals commonly used for this adjustment are sulfuric acid (H_2SO_4) and carbon dioxide (CO_2) to lower the pH and sodium carbonate (Na_2CO_3), lime ($Ca(OH)_2$) and caustic soda ($NaOH$) to raise the pH.

4.6.2.1. Influence of pH on aluminum-based coagulants

There is a limited pH range in which insoluble aluminum hydrates, such as aluminum hydroxide, are formed from a diluted alum solution. If the pH is too low, a more soluble compound, $Al(OH)_2^+$, is formed. The optimum pH range in a particular case depends on the water's physicochemical composition. Aluminum hydroxide precipitates are positively charged when the pH is lower than 8 and negatively charged in the opposite case. Similarly, soluble hydrolysis products are positively charged when the pH is acidic. The mean positive charge of the soluble reaction products of alum with water increases as pH decreases. If the pH is too high (above pH 8), aluminum complexes into $Al(OH)_4^-$. For soft and highly colored water, the optimal pH of $Al_2(SO_4)_3$ is often set between 6.0 and 6.5, but for turbid water with moderate alkalinity, it will probably be between 6.0 and 7.0. As the quality of raw water is variable in terms of turbidity, color, alkalinity and organic matter, the coagulation pH zone is kept between 6.0 and 6.5 to avoid high color or residual aluminum concentrations in the treated water. This is particularly true when alkalinity is high and/or when the raw water's pH is high. Significantly high doses of $Al_2(SO_4)_3$ are then necessary to reach the optimum pH for coagulation. The alternative is to dose the acid so that the pH is reduced to a lower value than before dosing the coagulant. On the other hand, alkalis (lime, sodium carbonate or caustic soda) are dosed to increase the pH of poorly alkaline waters.

Hydrogen ions will react to water alkalinity and, in doing so, will lower the water's pH, as shown in the equation for aluminum sulfate:

$$Al_2(SO_4)_3,18H_2O \rightarrow 2Al^{3+} + 3SO_4^{2-} + 18H_2O \rightarrow 2Al(OH)_3 + 6H^+ + 3SO_4^{2-} + 12H_2O$$

For the European standard, the limit residual aluminum concentration in treated water must be ≤ 0.2 mg·L^{-1}

Simplified solubility charts (Figure 4.20) have been developed to illustrate the relationships between metal salt concentrations and pH. Solubility charts can help describe the different species of metal salts present in varying pH waters. This information is particularly useful when choosing the coagulant's type and dosage.

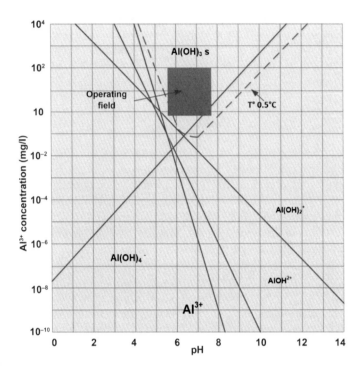

Figure 4.20. *Aluminum solubility chart as a function of pH (25°C).*
For a color version of this figure, see www.iste.co.uk/gaid/watertreatment1.zip

For a given coagulant concentration (expressed in Al^{3+}), the curves delimit the pH zone where the products of aluminum sulfate hydrolysis are the most available (in the form of a precipitate). Outside this zone, these products become soluble, and the coagulation mechanism is disrupted. A change in temperature results in a change in the curve position surrounding the amorphous aluminum hydroxide precipitate zone (dotted line in the figure). Figure 4.20 shows that as the temperature decreases, the minimum aluminum solubility pH increases.

Aluminum chloride (ACH) works in a pH range from 6.5 to 7.5. In some cases, the meaning of this could be that coagulation does not require an alkaline dosage. The following reaction takes place:

$$Al_2(OH)_5Cl \rightarrow Al_2(OH)_5{}^+ + Cl^- + H_2O \rightarrow 2Al(OH)_3 + H^+ + Cl^-$$

Only one mole of hydrogen ions is produced, reflecting the hydroxylated nature of this compound.

PACl also leads to hydrolysis, with a reaction showing that three moles of H^+ are formed:

$$Al_2(OH)_3Cl_3 \rightarrow Al_2(OH)_3{}^{3+} + 3Cl^- + 3H_2O \rightarrow 2Al(OH)_3 + 3H^+ + 3Cl^-$$

Hydrolysis reactions typically take place at a controlled water pH between 5.8 and 7.5, depending on the particular coagulant. When the pH value is below the isoelectric point of metal hydroxide (pH 7.7 for $Al(OH)_3$), it is the cationic complexes that act by charge neutralization. At this pH, color and colloidal matter are removed by adsorption onto and into the metal hydroxide hydrolysis products formed.

PolyDADMAC compounds perform quite well at any pH, which can be an advantage for some applications. However, most of the time, these are associated with a pH-controlling metallic coagulant.

4.6.2.2. Influence of pH on iron-based coagulants

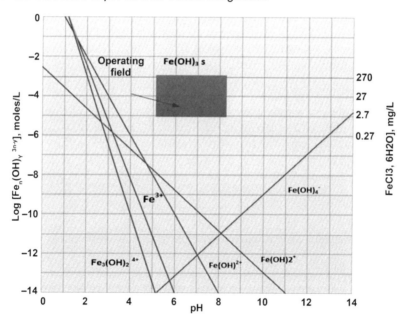

Figure 4.21. *Iron hydroxide equilibrium solubility chart at 25°C.*
For a color version of this figure, see www.iste.co.uk/gaid/watertreatment1.zip

When in solution, ferric salts immediately dissociate to form a hydrated reaction of the products. Metal ions form coordination compounds with water molecules and produce $\{Fe(H_2O)_6\}^{3+}$. The equation provides the global stoichiometric reaction of iron and shows that for each mole of trivalent metallic ion, one mole of amorphous (am) solid and three moles of hydrogen ions are produced. Iron acts as an acid because it releases hydrogen ions and lowers the water's pH:

$$Fe^{3+} + 3H_2O \rightarrow Fe(OH)_3, am + 3\ H^+$$

Ferric coagulants work properly over a wider pH range than aluminum. Therefore, its application in coagulation–flocculation is easier and less constraining than aluminum. The pH range for iron precipitation is between 5.0 and 8.0 (Figure 4.21), and its minimum solubility takes place at a pH of approximately 8.0. Ferric species are less soluble along a wider pH range. Positively charged hydrolysis products, as well as Fe^{3+} ions, are formed at low pH values, whereas only negatively charged hydrolysis products are formed at higher pH values.

One of the observations drawn from solubility charts is that iron-based coagulants have a wider pH range, in which precipitation is likely to occur, compared to that of aluminum. The curves surrounding the iron hydroxide precipitate zone show the balanced concentrations of soluble species in equilibrium with $Fe(OH)_3(s)$. For a given coagulant concentration (expressed in Fe^{3+}), these curves (Figure 4.21) delimit the pH zone where the products of ferric chloride hydrolysis are the most effective for coagulation and colloid destabilization.

When pH values are below the isoelectric point of metal hydroxide (pH 8 for $Fe(OH)_3$), it is the cationic complexes that act by charge neutralization. An excess of coagulant can destabilize the colloid by causing a charge inversion of the ions adsorbed on the particles' surface. This effect mostly occurs in the case of low turbidity. For pH values close to or above the isoelectric point, the main mechanism is the trapping and adsorption of colloidal particles by $Fe(OH)_3$. In practice, turbidity removal occurs at pH values between 6.8 and 8, where both mechanisms are complementary.

For ferrous sulfate, the precipitation equation for ferrous hydroxide is given as:

$$Fe^{2+} + 2OH^- \rightarrow Fe(OH)_2(s)$$

The precipitation of $Fe(OH)_2$ is observed in the pH range > 7. The precipitation pH of ferric hydroxide (pH 3) is lower than the precipitation pH of ferrous hydroxide. Thus, the domain of ferric hydroxide is larger than that of ferrous hydroxide. Its thermodynamic stability is superior.

4.6.3. Coagulation and flocculation times

Coagulant reactions are surprisingly quick. Aluminum sulfate $Al_2(SO_4)_3$ xH_2O hydrolyses as $Al(OH)_2$ in less than 10^{-3} s. A short and intense mixing of metallic salts is enough, especially when metallic salts are used for neutralizing particle charges (adsorption and destabilization). The formation of the aluminum hydroxide precipitate takes longer. Coagulation times in the frame from 1 to 2 min are recommended. Flocculation time requirements depend on temperature, floc concentrations, etc. Actually, the flocculation time varies from 10 to 30 min. If direct filtration follows flocculation, the time may be shorter (10–15 min). During settling, the time a particle stays in the reactor will affect the efficiency of its removal. The hydraulic retention time (or simply retention time) is defined by t:V/Q.

4.7. How to choose the best coagulant?

The choice of coagulant and the dose to be injected depend on the coagulants' characteristics (including its price), particle concentration and type, the characteristics of organic substances, water temperature and other raw water constituents, such as alkalinity. There is no formal approach on how to integrate this set of variables into the selection process. Feedback from fieldwork and jar tests play an important role in the selection process.

Water with high turbidity and high alkalinity is the easiest to coagulate. Aluminum sulfate, ferric chloride and high molecular weight polymers are regularly used successfully for this type of water. pH control is a crucial parameter for the coagulation of high turbidity and low alkalinity water. To regulate and control the coagulation pH, adding a base may be necessary for aluminum sulfate and ferric chloride. These two coagulants often require high doses for treating low turbidity and high alkalinity waters. These are the most difficult cases to coagulate because pH adjustment is usually necessary. Direct filtration should be considered for this type of water. Color coagulation is critically pH dependent. The flakes that form after the coagulation–flocculation of colored water are fragile because they are only metal hydroxides. The most common dosages are between 10 and 150 $mg \cdot L^{-1}$ for aluminum sulfate, 5 and 150 $mg \cdot L^{-1}$ for ferric chloride and 10 and 250 $mg \cdot L^{-1}$ for ferric sulfate.

The choice of coagulant greatly depends on the water to be treated (analysis of TOC, turbidity, pH, TH, TAC, iron, aluminum, manganese, calcium, chloride, etc.), water temperature (summer, winter), its calco-carbonic equilibrium and the treatment one wishes to obtain (removal of turbidity and/or TOC, color, etc.). For this reason, it is necessary to carry out jar tests to find the right coagulant for the water to be treated.

The parameters to know beforehand are as follows:

– Basicity: this gives an indication of the number of hydroxyl ions included in the structure of a polymerized coagulant. The higher this is, the lower the impact in relation to pH. For example, aluminum hydrochloride $Al_2(OH)_5Cl$ at 83–85% basicity shows less impact on pH than another polymerized coagulant only having three OH ions in the structure and therefore a 50–55% basicity. Aluminum sulfate is not polymerized and has no OH ions in its structure, thus having no basicity.

– Percentage Al_2O_3 for aluminum-based coagulants: For example, aluminum sulfate ($Al_2(SO_4)_3,14H_2O$) is present in liquid form and typically contains 8.5% Al_2O_3. This corresponds to 4.6% aluminum weight.

– For ACH and PACl coagulants, the brand name often includes the Al_2O_3 percentage. For example, PAC-23 produced from aluminates is an ACH with a nominal Al_2O_3 content of 23.5%, that is, 12.7% Al weight. This is equivalent to ACH with a concentration of 40.2% $Al_2(OH)_5Cl$ weight.

– Similarly, a PACl with a nominal Al_2O_3 content of 10.5% corresponds to 5.6% Al or 21.8% $Al_2(OH)_3Cl_3$.

– Density is often expressed in $kg \cdot L^{-1}$. As an example, liquid $Al_2(SO_4)_3$, $14H_2O$ (49% w/w) typically has a density of 1.32.

Calculation of the coagulant dose ($mg \cdot L^{-1}$) is indicated as follows.

Consider an aluminum-based coagulant $Al_2(SO_4)_3,14H_2O$, that is, 594 $g \cdot mole^{-1}$ with 8.5% Al_2O_3 content and a density equal to 1.32.

This is equivalent to $594 \times 8.5/100 = 50.5\%$ w/w $Al_2(SO_4)_3,14H_2O$.

The coagulant's concentration is therefore $10,000 \times 50.5 \times 1.32 = 666,000$ $mg \cdot L^{-1}$ or 666 $g \cdot L^{-1}$.

If the coagulant is dosed at a rate of 150 $mL \cdot min^{-1}$ and the raw water flow is 50 Ls^{-1}, then the coagulant dose is equal to

$$(150 \times 666,000/(1,000 \times 60))/50 = 33.3 \ mg \cdot L^{-1}$$

In general, aluminum sulfate is chosen due to its lower cost and availability. For colored, low turbidity and low pH/alkalinity surface waters, CO_2–lime preremineralization is normally required to obtain the optimum pH for coagulation. Sometimes only a pretreatment with lime, sodium or caustic soda is implemented, but this procedure does not stabilize pH due to variations in the quality of raw water, which requires continual adjustments in the dosage of coagulants and added alkalis.

However, these types of water can be treated with polyaluminum chlorides (ACH, PAC) or polychlorosulfates and/or aluminum silicates (PASS, PAX, etc.), because these products allow coagulation at a relatively high pH (7.5–7.8), consume little alkalinity and thus avoid the need to dose an alkali base for pH correction.

Iron-based coagulants, such as ferric chloride, ferric sulfate and ferrous sulfate, are particularly efficient for heavily loaded waters (color and organic matter). However, they consume more alkalinity than Al-based coagulants and therefore tend to lower the pH more evidently.

Organic coagulants such as the polyDADMAC liquid cationic polymers are not as effective as inorganic coagulants for removing true color and NOM from water. Often, when a polyDADMAC organic coagulant is combined with aluminum sulfate or with polyaluminum chloride, the total dose of chemical required to obtain the same final water quality can be lower than if each chemical had been used separately. This combination with a polyDADMAC cationic polymer can be effective on highly colored waters with low alkalinity. However, an excess dosage of polyDADMAC cationic polymers can lead to a colloidal particle restabilization phenomenon, which is explained by a modification in the surface charge of colloidal particles, changing from negative to positive. Water turbidity increases.

4.7.1. *How to choose the optimal dose of coagulant?*

Cost and performance are the criteria to be optimized. The optimum dosage can be defined as the lowest dose that produces an easily decantable floc to remove turbidity in a reasonably short time, as well as remove excess water color and organic matter.

Figure 4.22 presents testing feedback on Al^{3+} dosages obtained in waters with variations in color and turbidity.

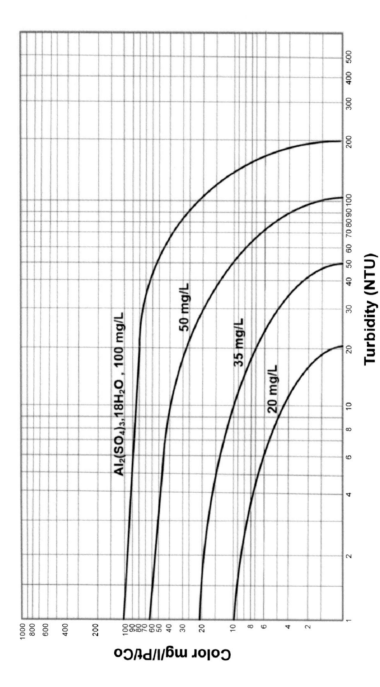

Figure 4.22. *Removal of turbidity and color with $Al_2(SO_4)_3,18H_2O$*

4.8. Residual aluminum

Residual aluminum in treated water must be as low as possible and, in any case, must respect the 0.2 mg·L^{-1} standard. When the water leaving the treatment facilities contains a high aluminum concentration (greater than 0.4 mg·L^{-1}), spontaneous flocculation may occur within the distribution network. Flocs will make the water more turbid. This is why the choice of Al-based coagulant is crucial when operating under conditions that minimize residuals in the water (type of coagulant, pH, alkalinity, dosage, etc.).

Figure 4.23A illustrates the experimental relationship obtained between residual aluminum and the treatment pH. However, the optimum pH that is needed to minimize residual aluminum also depends on other substances present in the solution. For example, in an Al^{3+}, pH and alkalinity matrix, it has been observed that when the concentration in alkalinity (mg·L^{-1} CaCO$_3$) increases, residual aluminum tends to increase as well (Figure 4.23B). Other parameters can have an impact, such as the presence of fluorides (>5 mg·L^{-1}) in raw water, which induces a higher pH in minimal residual aluminum (pH shifts to approximately 7). In the presence of NOM, the complexation of aluminum species with humic substances can generate a high residual Al.

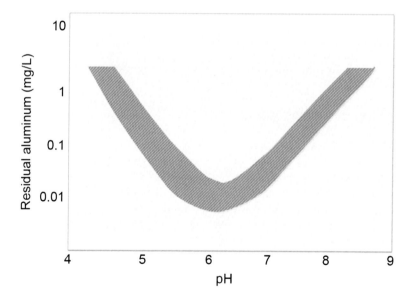

Figure 4.23A. *Al^{3+} residual versus pH and experimental curve obtained on site*

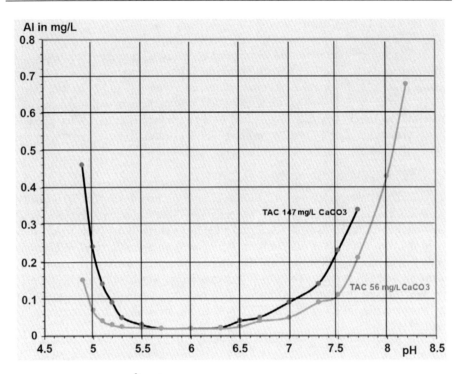

Figure 4.23B. Al^{3+} *residual versus pH and experimental curve obtained on site, with an alkalinity variation. For a color version of this figure, see www.iste.co.uk/gaid/watertreatment1.zip*

4.9. Alkalinity consumption

Alkalinity is a major economic parameter for the removal of organic compounds because, together with the type of coagulant and dosage, it determines the pH that can be obtained without resorting to an acid or a base, either during the pretreatment or later adjustment. When a metal salt is used without further addition of an acid or a base, the pH is controlled only by virtue of the coagulant's dosage. However, this system is not flexible. The effects of alkalinity in relation to metallic salts such as ferric chloride and aluminum sulfate are described below.

Coagulants based on trivalent metals have several effects on the characteristics of coagulated water. The action of a coagulant (aluminum or iron-based) causes a drop in pH and TAC because the coagulant – which is a strong acid salt – displaces CO_2 hydrogen carbonates, which are weak acid salts. There is therefore an increase in free CO_2, which lowers the pH. This is problematic insofar as pH values influence coagulation and colloid flocculation. For this, it is necessary to correct the pH upstream to enable the raw water to arrive at the coagulation/flocculation stage with the optimum pH value for coagulation/flocculation. In cases of low alkalinity, the pH may be low enough to enable the removal of several organic compounds. However, in waters containing few colloids and particles, coprecipitation is difficult and makes this approach less efficient. The coagulation of dissolved organic matter benefits from the addition of particles (adjuvants), an increase in alkalinity or a combination of these two factors. In the case of high-alkalinity water, the coagulation–flocculation pH may still be too high despite the coagulant dosage decided upstream to eliminate turbidity and organic compounds. A first solution may be to increase the coagulant dose to consume alkalinity and thus reach the optimum pH. A second solution could be to control the pH by adding an acid while maintaining the previously determined coagulant dose.

The chemical reactions of the two main coagulants are as follows.

4.9.1. *Aluminum and alkalinity*

The reaction between aluminum sulfate and alkalinity is:

$$Al_2(SO_4)_3, 14\ H_2O + 6HCO_3^- \rightarrow 2\ Al(OH)_3, 3H_2O + 6CO_2 + 8H_2O + 3\ SO_4^{2-}$$

$$\text{or } Al_2(SO_4)_3 + 6HCO_3^- \rightarrow 2\ Al(OH)_3 + 6CO_2 + 3\ SO_4^{2-}$$

Each mole of added Al sulfate consumes six moles of alkalinity and produces six moles of CO_2. We obtain a decrease in pH due to the CO_2 produced. As long as water alkalinity is sufficient, the production of CO_2 lowers the pH, but this still remains within acceptable operating conditions. If, on the other hand, alkalinity is not sufficient to neutralize the production of sulfuric acid, then the pH drops very sharply (Table 4.6). In this case, (quick or slaked) lime, soda or sodium carbonate must be added to neutralize the acidity formed. Otherwise, the $Al(OH)_3$ precipitate dissolves back again:

$$Al_2(SO_4)_3, 14\ H_2O \rightarrow 2\ Al(OH)_3, 3H_2O + 5H_2O + 3H_2SO_4$$

For example, the amount of alkalinity consumed after the addition of 100 mg·L^{-1} of aluminum sulfate is given as follows:

The number of added Al sulfate moles is given as:

$$\frac{100\ \text{mg·L}^{-1}}{1{,}000\ \text{mg·L}^{-1}\ 594\ \text{mole·L}^{-1}} = 1.68 \times 10^{-4}\text{mole·L}^{-1}$$

$$M\ Al_2(SO_4)_3, 14\ H_2O = 594\ \text{g·mole}^{-1}$$

The alkalinity consumed is given as

$$6 \times 1.68 \times 10^{-4} = 1.01 \times 10^{-3} \text{ mole·L}^{-1} \times 61 = 0.0616 \text{ g·L}^{-1} = 61.6 \text{ mg·L}^{-1}$$
HCO_3^-, equivalent to 50.5 mg·L^{-1} $CaCO_3$

The general rule used is that 1 mg·L^{-1} of Al sulfate consumes 0.5 mg·L^{-1} of alkalinity (expressed as $CaCO_3$).

Since 3 moles of H_2SO_4 are produced for each mole of Al sulfate, we obtain

$$3 \times 1.38 \times 10^{-4} = 5.04 \times 10^{-4} \text{ mole·L}^{-1} H_2SO_4$$

with the dissociation of H_2SO_4:

$$H_2SO_4 \rightarrow 2H^+ + SO_4^{2-}$$

That is:

$$[H^+] = 2 \times 5.04 \times 10^{-4} \text{ mole·L}^{-1} = 1.1 \times 10^{-3} \text{ mole·L}^{-1}$$

$$pH = -\log [H+] = -\log (1.1 \times 10^{-3}) = 3$$

The amount of caustic soda needed to neutralize the pH is calculated as:

$$H_2SO_4 + 2NaOH \rightarrow Na_2SO_4 + 2H_2O$$

Two moles of NaOH are needed for 1 mole of H_2SO_4, that is;

$$2 \times 1.1 \times 10^{-3} = 2.02 \times 10^{-3} \text{ mole·L}^{-1} \text{ sodium} = 2.02 \times 10^{-3} \times 40 \text{ g·mole}^{-1} \times$$
$$1{,}000 \text{ mg·g}^{-1} = 81 \text{ mg·L}^{-1}$$

To determine whether a base should be added in the presence of alkalinity in the water, this should be calculated following the previous example and deducting the proportion that will be consumed by the coagulant's acidity. If the proportion consumed exceeds the amount present in the water, it will then be necessary to estimate the quantity of base to be added to neutralize acidity.

The reaction equation shows that one mole of $Al_2(SO_4)_3$ reacts with 3 moles of $Ca(HCO_3)_2$:

$$Al_2(SO_4)_3 + 3Ca (HCO_3)_2 \rightarrow 2Al (OH)_3(s) + 3CaSO_4 + 6CO_2$$

Available presentations	Aluminum sulfate
Hydrolysis reaction	$Al_2(SO_4)_3 + 6\,H_2O \rightarrow 2Al(OH)_3(s) + 3H_2\,SO_4$
Influence on TAC	$3H_2SO_4 + 3Ca(HCO_3)_2 \rightarrow 3CaSO_4 + 6CO_2 + 6H_2O$
Overall reaction	$Al_2(SO_4)_3 + 3Ca\,(HCO_3)_2 \rightarrow 2Al\,(OH)_3(s) + 3CaSO_4 + 6CO_2$

Table 4.6. *Chemical reactions showing the consumption of alkalinity by aluminum sulfate*

With an $Al_2(SO_4)_3$ coagulant at 17.2% in Al_2O_3 (basicity 0.3%), the Al concentration (mole L^{-1}) is given as:

$$3.37 \times 10^{-6} = \frac{1}{1,000}\frac{17.2}{100}\frac{2}{102}$$

M $Al_2O_3 = 102$ g·mole^{-1}

The HCO_3^- (mole·L^{-1}) consumption is

$$1.10 \times 10^{-5} = 3\left(1 - \frac{0.3}{100}\right)3.37 \times 10^{-6}$$

That is:

$1.10 \times 10^{-5} \times 50,000 = 0.50$ mg $CaCO_3$·mg^{-1} Al^{3+}.

Following an equivalent calculation, with an Al_2O_3 content at 8.3% (basicity 0.5), we obtain a consumption of 0.24 mg $CaCO_3$/mg coagulant.

Another method can be used and based on the chemical reaction and the molar consumption between aluminum and carbonates:

$$Al_2(SO_4)_3 + 3Ca\,(HCO_3)_2 \longrightarrow 2Al\,(OH)_3(s) + 3CaSO_4 + 6CO_2$$

One mole of $Al_2(SO_4)_3$ reacts with 3 moles of Ca $(HCO_3)_2$.

One of $Al_2(SO_4)_3$, 14 H_2O corresponds to 594 g.

Three moles of $CaCO_3$ correspond to 300 g.

This is a ratio (300/594) in the range of 0.50 mg $CaCO_3$ consumed per milligram of injected $Al_2(SO_4)_3$ 14 H_2O.

The consumption of $Ca(HCO_3)_2$ is expressed in $CaCO_3$.

The pH is lowered due to this alkalinity consumption. This hypothesis is verified in practice by the appearance of $Al(OH)_3(s)$ precipitates. However, the addition of aluminum sulfate transforms temporary hardness (calcium bicarbonate) into permanent hardness (calcium sulfate). This means that there is a release of carbon dioxide, making the water acidic and corrosive.

For example, if 30 mg·L^{-1} $Al_2(SO_4)_3$ 14 H_2O (at 17.2%) is introduced into water containing an alkalinity of 20 mg·L^{-1} $CaCO_3$ and pH 7.2, its pH will drop to 5.8, and the alkalinity drops to 15 mg·L^{-1} $CaCO_3$. Therefore, the coagulation pH needs to be adjusted. Thus, preremineralization is required (increasing alkalinity) to operate within the optimum pH range.

4.9.2. Iron and alkalinity

The reactions are presented in Table 4.7.

Available presentations	$FeCl_3$
Hydrolysis reaction	$2\ FeCl_3 + 6H_2O \rightarrow 2\ Fe\ (OH)_3(s) + 6HCl$
Influence on alkalinity	$6HCl + 3Ca\ (HCO_3)_2 \rightarrow 3CaCl_2 + 6CO_2 + 6H_2O$
Overall reaction	$2FeCl_3 + 3Ca\ (HCO_3)_2 \rightarrow 2Fe\ (OH)_3(s) + 3CaCl_2 + 6CO_2$

Table 4.7. *Chemical reactions showing the consumption of alkalinity by ferric chloride*

Two moles of pure $FeCl_3$ react with three moles of $Ca(HCO_3)_2$. That is, for 2 moles of pure $FeCl_3$ (which correspond to 326 g), the consumption of 3 moles of $CaCO_3$ is 300 g. That is, a ratio of 0.92 g $CaCO_3$ was consumed per gram of (pure) injected $FeCl_3$. Nevertheless, 1 mg of pure $FeCl_3$ consumes 0.92 mg·L^{-1} $CaCO_3$.

If the $FeCl_3$ coagulant used is 41%, the alkalinity consumption becomes equal to $0.92 \times 0.41 = 0.38$ mg $CaCO_3$ consumed/mg $FeCl_3$ at 41%. The pH is lowered due to this alkalinity consumption. This hypothesis is verified in practice by the appearance of $Fe(OH)_3(s)$ precipitates.

The reaction with ferric sulfate is given as:

$$Fe_2(SO_4)_3,\ 9H_2O + 3Ca\ (HCO_3)_2 \rightarrow 2Fe\ (OH)_3 + 3CaSO_4 + 6CO_2 + 9H_2O$$

Pure ferric sulfate (1 g) consumes 0.75 g of alkalinity. One milligram of pure ferric sulfate consumed $0.075°F$ (0.75 mg·L^{-1} CaCO$_3$). A similar calculation leads to a consumption of $0.038°F$/mg of Fe$_2$(SO$_4$)$_3$ at 43% or 0.38 mg CaCO$_3$/mg of Fe$_2$(SO$_4$)$_n$.

For polymerized coagulants, the following formula takes basicity into account:

$$\Delta alk.\,consumed = \frac{300 \times 3 \cdot Basicity}{2,700} \quad meq\cdot mg^{-1}\,Al$$

Thus, for example, a coagulant X with 75% basicity will consume 0.028 meq.mg^{-1} of alkalinity. Table 4.8 and Figure 4.24 summarize the alkalinity consumption of various coagulants.

Depending on water alkalinity, that is, its buffering capacity, the addition of an acidic or alkaline chemical may be required to adjust the pH in the optimal coagulation zone. The impact of coagulants on alkalinity and pH can be corrected by adding alkalis such as lime. CO$_2$ neutralization by lime is expressed as:

$$2CO_2 + CaO + H_2O \;\rightarrow\; Ca(HCO_3)_2$$

This pH adjustment is carried out by adding various bases:

$$Al_2(SO_4)_3\,14(H_2O) + 6OH^- \rightarrow 2Al(OH)_3 + 3SO_4^{2-} + 14H_2O$$

$$Al_2(SO_4)_3\,14(H_2O) + 6Ca\,(HCO_3)_2 \rightarrow 2Al(OH)_3 + 3CaSO_4 + 6CO_2 + 14H_2O$$

$$Al_2(SO_4)_3\,14(H_2O) + 6NaHCO_3 \rightarrow 2Al(OH)_3 + 3Na_2SO_4 + 6CO_2 + 14H_2O$$

$$Al_2(SO_4)_3\,14(H_2O) + 3Na_2CO_3 \rightarrow 2Al(OH)_3 + 3Na_2SO_4 + 3CO_2 + 14H_2O$$

$$Al_2(SO_4)_3\,14(H_2O) + 6NaOH \rightarrow 2Al(OH)_3 + 3Na_2SO_4 + 14H_2O$$

$$Al_2(SO_4)_3\,14(H_2O) + 3Ca\,(OH)_2 \rightarrow 2Al(OH)_3 + 3CaSO_4 + 14H_2O$$

For 1 mole of an Al$_2$(SO$_4$)$_3$ 14(H$_2$O) commercial solution, 3 moles of pure lime are needed to neutralize the 3 moles of H$_2$SO$_4$. Thus, 0.37 g of lime is necessary to compensate for the acidity caused by the addition of 1 g Al$_2$(SO$_4$)$_3$ 14H$_2$O.

Products	Formulae	Commercial product (% mm^{-1})	Density (kg·m^{-3})	HCO$_3^-$ consumed (°f·g^{-1})	HCO$_3^-$ consumed (gCaCO$_3$·g^{-1})	gCaCO$_3$·g^{-1} pure coagulant
Ferric chloride	FeCl$_3$, 6 H$_2$O	41%	1,450	0.038	0.38	0.92
Ferric sulfate	Fe$_2$(SO$_4$)$_3$, 9 H$_2$O	43%	1,500	0.032	0.32	0.75
Ferrous sulfate	FeSO$_4$,7H$_2$O	28%	1,500	0.018	0.18	0.66
Aluminum sulfate (liq)	Al$_2$(SO$_4$)$_3$, 18 H$_2$O	8.5% Al$_2$O$_3$	1,330	0.024	0.24	0.51
Aluminum sulfate (s)	Al$_2$(SO$_4$)$_3$, 14 H$_2$O	17% Al$_2$O$_3$	1,690	0.050	0.50	0.50
WAC	[Aln(OH)mCl$_3$m-n-n]x	100%	1,200	0.0147	0.147	0.147
HB-WAC	[Aln(OH)mSO$_4$kCl$_3$m-n-2k]x	100%	1,170	0.008	0.08	0.075
WAC HBA	[Aln(OH)mSO$_4$kCl$_3$m-n-2k]x	100%	1,200	0.008	0.08	0.08
Aqualenc	[Aln(OH)mSO$_4$kCl$_3$m-n-2k]x	100%	1,160	0.012	0.12	0.12

Products	Formulae	Commercial product (% mm^{-1})	Density (kg·m^{-3})	HCO$_3^-$ consumed (°f·g^{-1})	HCO$_3^-$ consumed (gCaCO$_3$·g^{-1})	gCaCO$_3$·g^{-1} pure coagulant
Aqualenc F1	[Aln(OH)mSO$_4$kCl$_3$m-n-2k]x	100%	1,240	0.009	0.09	0.09
Aqualenc E	-	100%	1,370	0.031	0.31	0.31
PACl	Al$_2$(OH)$_3$Cl SO$_4$	100%	-	0.057	0.57	0.57
PAX 14	-	100%	1,300	0.025	0.25	0.25
PAX 18	[Aln(OH)mCl$_3$m-n]x	100%	1,360	0.029	0.29	0.29
PAX 20	-	100%	1,280	0.020	0.20	0.20
PAX XL31	-	100%	1,350	0.027	0.27	0.27
PAX XL60	-	100%	1,310	0.013	0.13	0.13
PAX XL9	[Aln(OH)mSO$_4$kCl$_3$m-n-2k]x	100%	1,200	0.008	0.08	0.08
PAX XL10	[Aln(OH)mSO$_4$kCl$_3$m-n-2k]x	100%	1,220	0.009	0.09	0.09
PAX XL19; ACH	Al$_2$(OH)$_5$Cl	100%	-	0.029	0.29	0.29
PAX XL63	[Aln(OH)mSO$_4$kCl$_3$m-n-2k]x	100%	1,220	0.010	0.10	0.10
PAX XL7A	-	100%	-	0.009	0.09	0.09

Table 4.8. *Alkalinity consumption by various coagulants*

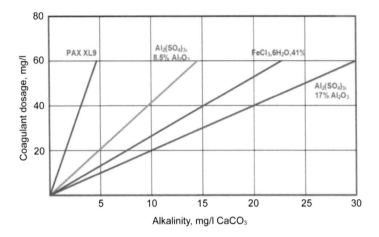

Figure 4.24. *Alkalinity consumption by some coagulants.
For a color version of this figure, see www.iste.co.uk/gaid/watertreatment1.zip*

4.9.2.1. *Neutralization reactions*

The neutralization reactions are as follows:

– With sulfuric acid:

- Lime: $H_2SO_4 + CaO \rightarrow CaSO_4 + H_2O$

$H_2SO_4 + Ca(OH)_2 \rightarrow CaSO_4 + 2H_2O$

- Caustic soda: $H_2SO_4 + 2NaOH \rightarrow Na_2SO_4 + 2H_2O$

- Sodium carbonate: $H_2SO_4 + Na_2CO_3 \rightarrow Na_2SO_4 + H_2O + CO_2$

– With hydrochloric acid:

- Lime: $2HCl + CaO \rightarrow CaCl_2 + H_2O$

$2HCl + Ca(OH)_2 \rightarrow CaCl_2 + 2H_2O$

- Caustic soda: $HCl + NaOH \rightarrow NaCl + H_2O$

- Sodium carbonate: $2HCl + Na_2CO_3 \rightarrow 2NaCl + H_2O + CO_2$

4.9.2.2. *The importance of water mineralization for the coagulation stage*

Total alkalinity indicates the content of carbonates, hydrogen carbonates and free bases of calcium, magnesium and sodium:

Total alkalinity $= [HCO_3^-] + [CO_3^{2-}] + [OH^-]$

Total alkalinity provides valuable information about water mineralization. It is an essential parameter to take into account during coagulation. Its evolution makes it possible to monitor pH correction. Preremineralization is mandatory in fresh water to compensate for the drop in total alkalinity and pH following the introduction of a coagulant. It is done by adding a base (lime or sodium) and CO_2.

The preremineralization reactions are as follows:

– with lime:

$$2CO_2 + CaO + H_2O \ \rightarrow \ Ca(HCO_3)_2$$

– with caustic soda:

$$CO_2 + NaOH \ \rightarrow \ NaHCO_3$$

$$NaHCO_3 + NaOH \ \rightarrow \ Na_2 CO_3 + H_2O$$

Since lime is cheaper and more readily available, it is often used in the field. In addition, and unlike caustic soda, it increases THCa. Adding caustic soda or lime makes it possible to compensate for the drop in pH, but this addition does not modify water remineralization, except in the presence of CO_2. For this reason, it is necessary to ensure that CO_2 is present at a sufficient concentration in raw water. Otherwise, it is necessary to add more, which actually occurs in the majority of cases so that chemical reactions can take place properly. If the additional supply of CO_2 were lacking (bearing in mind that its initial concentration may vary at any time in raw water), a simple addition of lime or caustic soda would not create the buffer effect provided by bicarbonates. In this case, the acidity delivered by the coagulant would not be sufficiently neutralized by lime or caustic soda. A rapid drop in pH was then observed. Coagulation would be uncontrolled.

4.10. Reduction efficiency of some water constituents

4.10.1. Turbidity and SS

Generally, an increase in the concentration of turbidity/suspended matter in the water requires an increase in the coagulant dosage to obtain an optimal reduction. In addition, pH influences coagulation, and a coagulant dose/pH trade-off must be chosen to obtain the lowest turbidity (Figure 4.25).

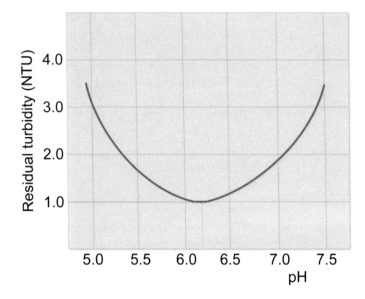

Figure 4.25. *Example of turbidity removal with aluminum sulfate (15°C) (surface water turbidity: 12 NTU, coagulant: 20 mg·L^{-1}, polymer flocculant: 0.1 mg·L^{-1})*

As turbidity increases, the coagulant concentration should also be increased. Nevertheless, the increase in the amount of coagulant does not vary linearly with the increase in turbidity. When turbidity is very high, it is possible for the coagulant dosage to be lower than expected because the probability of interparticle collisions is also very high. On the other hand, when turbidity is low, coagulation is difficult. In view of this, it is vital to choose the right coagulant and the right operating conditions. The particulate organic matter does not have any influence on the required coagulant dosage.

4.10.1.1. *Excess coagulant*

Excessive ion adsorption can cause a reversal in the zeta potential, which is responsible for the hydrodynamic behavior of colloids. In that case, there would be restabilization of the colloids in the solution. In fact, two colloids having overadsorbed cations from the coagulant tend to move away from one another.

Many models are available for calculating coagulant dosage correlations versus turbidity. The example in Figure 4.26 shows surface water with variable turbidity and a low concentration of dissolved organic matter (DOC).

The following ratio is obtained:

$Al_2(SO_4)_3$ (mg·L^{-1}) dosage = 7.8 (turbidity)$^{0.4}$

With a commercial solution of $Al_2(SO_4)_3$, 18 H_2O, the relation becomes:

$Al_2(SO_4)_3$, 18 H_2O, mg·L^{-1} dosage = 3.51 [(7.8(turbidity)$^{0.4}$]

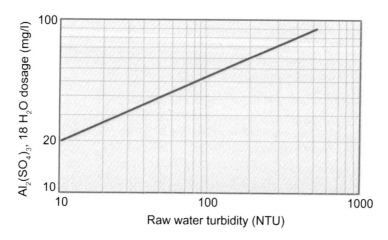

Figure 4.26. *Example of an empirical correlation model between Al sulfate dosage and raw water turbidity*

Table 4.9 gives some indications on the different dosages of Al sulfate and ferric chloride applied depending on the quality of raw water.

Dosage (gm^{-3})	Aluminum sulfate Crystalized $Al_2(SO_4)_3$, 18H_2O	Ferric chloride (FeCl$_3$ 41%)
Flocculation on filter	3–10	1.5–8
Settling		
Lightly loaded water	15–30	6–12
Medium turbidity waters	30–60	20–50
Highly turbid waters	60–150	50–100
Colored to very colored waters	60–250	50–350
Plankton-rich waters	60–150	25–100

Table 4.9. *Usual treatment rates for coagulation*

4.10.2. *Microorganism removal*

When removing turbidity and/or organic matter or algae, coagulation and flocculation operating conditions must be optimal: it is imperative to control the coagulant dose, the pH and quality of the water to be treated, as well as to arrange the appropriate mixing conditions for flocculation to take place. In this case, a reduction in microorganisms (bacteria, viruses and parasites) is obtained after liquid/solid separation (settling or flotation), as shown in Figure 4.27. However, if the operating conditions are not optimal, this will result in poor turbidity removal, which will lead to the observation of microorganisms.

4.10.2.1. *The case of bacteria (0.5–3 μm)*

Both aluminum and iron-based coagulants encourage the reduction of at least 2 log (or even 3 log) for all classes of pathogenic agents of bacterial origin. On the other hand, less than 1 log can be obtained if the operating conditions are not optimal, and if, for example, a turbidity leak is detected.

4.10.2.2. *The case of parasites (Cryptosporidium (3–6 μm) and Giardia (7–12 μm))*

Optimal coagulation–flocculation–settling conditions are governed by the removal of turbidity and organic matter (TOC) rather than by the removal of pathogens. These are the consequence. Properly operated drinking water plants with coagulation, flocculation, settling and filtration facilities are given a 2.5–2.8 log removal credit for *Cryptosporidium* and *Giardia*, often leaving only 0.5 log inactivation to be treated by disinfection when a 3 log reduction is needed. For a 5–6 log reduction across the entire process, membrane filtration (microfiltration or ultrafiltration) or UV disinfection (3 log) at the end of treatment is advisable. Coagulation and flocculation, together with clarification by dissolved air flotation (Spidflow, Veolia), make it possible to obtain between 1.8 and 2.2 log reduction of these parasites. These results are explained by the fact that the surfaces of *Cryptosporidium* oocysts are made up of layers of polysaccharides and that the negative charge carried by oocysts comes from the carboxylic acid groups in the surface proteins. Parasite removal by mineral coagulants (Al^{3+}) appears to occur by a sweep floc mechanism. In the presence of poor organic matter (DOC) concentrations, zeta potential measurements suggest that removal does not appear to occur via a charge-neutralizing mechanism. For high DOC concentrations, the mechanism is based on bridging between organic matter-aluminum hydroxide-oocyst particles.

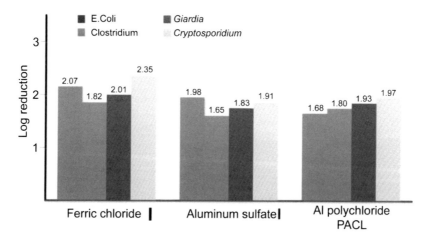

Figure 4.27. *Bacteria and parasite reduction. For a color version of this figure, see www.iste.co.uk/gaid/watertreatment1.zip*

4.10.2.3. *The case of viruses (20–100 nm)*

Significant virus (>3 log) reductions are obtained using metal coagulants and polyelectrolytes. Although the use of polyelectrolytes as flocculants improves floc formation, it does not help with virus suppression beyond the levels removed with metallic coagulants alone. Viruses are essentially DNA (deoxyribonucleic acid) or RNA (ribonucleic acid) units contained within a protein coat. The destabilization mechanism involves coordination reactions between the metal coagulant species and the carboxyl groups of the virus's protein coat. Due to the similarity between the destabilization mechanisms of organic color and viruses, reductions are high when pH values are acidic and in the presence of high coagulant doses. Metal coagulants do not completely inactivate viruses, and the final disinfection of water before distribution is required to achieve 5-6 log reductions. Organic coagulants of the PACl type have a higher virucidal activity than Al sulfate. Bacteriophage MS2 and human enteric polioviruses are removed with fairly high efficiency (2.6–3.4 log), whereas PRD-1 phage and the enteric Echovirus are removed with lower efficiency (1.1–1.9 log). Differences in virus removal were significant for Al sulfate (Figure 4.28). The efficiency of coagulation is different from one virus to another, and it is not advisable to extrapolate data relating to known viruses compared to other untested viruses.

Some factors that may cause poor efficiency when removing microorganisms by coagulation–flocculation and settling are variable flow rates for which the facilities are not prepared, inadequate coagulant doses, unsatisfactory control of operating parameters, excessive agitation (which provokes floc shearing), an inappropriate mixture of chemical products, and uncontrolled sewage sludge extraction.

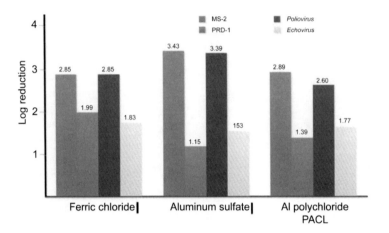

Figure 4.28. *Reduction of some viruses. For a color version of this figure, see www.iste.co.uk/gaid/watertreatment1.zip*

4.11. Jar tests

The efficiency of coagulation–flocculation depends on many variables such as:

– type of coagulant used;

– coagulant dosage;

– final pH;

– coagulant feed concentration;

– type and dosage of chemical additives, other than the primary coagulant (e.g. polymers);

– chemical addition sequence and time lag between dosage applications;

– mixing intensity and duration at the rapid mixing stage;

– type of rapid mixing device;

– velocity gradients applied during flocculation;

– flocculator retention time;

– type of stirrer used.

The best approach to determining the treatability of a water source and the optimum parameters is to perform a jar test. The main interest of the jar test lies in the information they offer about the best coagulation–flocculation operating conditions, as well as the possibility of observing the behavior of suspensions not only within an optimum but also on either side of it. The normal procedure when performing the jar test is to first choose the best coagulant, then the flocculant (as well as the associated dosages that will be the most effective), and finally determine the optimum pH. Performance is generally assessed regarding turbidity and color or organic matter removal when these parameters are required for the quality of the treated water. Moreover, a pH correction may be needed. The jar-test protocol implies that each sample undergoes the same rapid mixing for coagulation, followed by slow agitation for flocculation. The jar tester is a device equipped with six 1-L beakers, an agitator with blades or propellers for each beaker, and a variable velocity engine (30–230 rpm^{-1}) for rotational steering. Agitation engines are subject to a clock for parametrizing agitation time (Figure 4.29).

One liter of water was placed in each of the six beakers. Then, variable coagulant dosages are introduced into the various beakers. Rapid agitation (approximately 100 rpm^{-1}) was initiated for 2 min. This simulates the coagulation stage. At the end of this period, the flocculant is introduced into each beaker, and slow agitation is initiated (<50 rpm^{-1}) for 10–20 min.

The value of G can be estimated using the following ratio:

$$G = \left(\frac{D^5 k}{V\eta}\right)^{0.5} (N^{1.5})$$

where:

– G: velocity gradient (s^{-1});

– D: blade length or diameter (m);

– k: blade constant characteristic, varying between 0.1 and 5;

– V: beaker volume (1 L or 0.001 m^3);

– η: kinematic viscosity (1.1 × 10^{-6} m^2·s^{-1}).

During these two stages, the flocs obtained are observed and their dimensions are noted. After floc settling (approximately 30 min), the parameters sought were analyzed on the supernatant. This is used to proceed with the optimal conditions setting, such as the best pH conditions for coagulation by adding an acid (sulfuric acid) or a base (lime, caustic soda). This makes it possible to find the optimum pH leading to the best coagulation result and to obtain the lowest possible residual metal.

The optimal reagent dose chosen corresponds to the one for which the best efficiency is observed using a minimum concentration of reagents. The choice of coagulant and flocculant also takes into account the reaction time, floc settling time and floc volume.

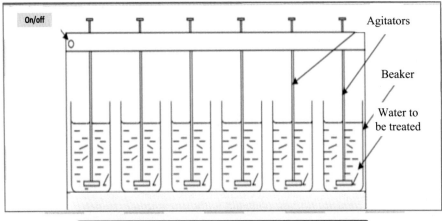

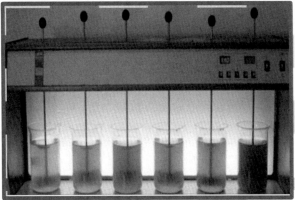

Figure 4.29. *Jar test apparatus. For a color version of this figure, see www.iste.co.uk/gaid/watertreatment1.zip*

Theoretically, the results of a jar-test protocol should help in designing coagulation–flocculation–settling structures and provide valuable information concerning the removal performance of the desired parameters. However, the actual performance of the drinking water plant, in terms of clarification, settling rate or other parameters, may not be identical to the ones found during testing. It is still possible to establish an empirical relationship that can be used for predictive purposes. The mixing intensity at the laboratory is kept constant during the two coagulation/flocculation stages, whereas variations are often observed at the plant level. In addition, settling in a beaker is static, and flocs are not subject to the dynamics provided by the rising velocity. On the other hand, the lamellae present in lamellar settling tanks such as Multiflo® or Actiflo® compensate for the settling static effect in the laboratory.

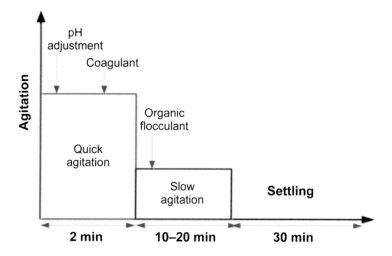

Figure 4.30. *Agitation modes during jar tests*

In general, measuring three parameters (sedimentation rate, settled volume and supernatant turbidity or TOC) provides enough information about the treatability conditions of raw water. The measurement of supernatant turbidity is frequently used as an indicator of flocculation performance.

Overdosing coagulants and/or flocculants and disregarding agitation conditions or contact times may lead to a restabilization of some of the particles, as well as to an increase in the supernatant liquid's residual turbidity (Figure 4.31). The jar test is one of the most important tools an operator possesses to evaluate and optimize different chemical dosing regimes, as well as to regularly monitor the impact of his/her actions. Despite the compensation effects of lamellae, a better removal at the laboratory is obtained compared to the one recorded at the plant.

When turbidity is the only parameter to be removed, coagulants are chosen depending on their local availability and cost. The most often used is aluminum sulfate (18 H_2O). When available, organic coagulants (PAC or PACl, PACS, PASS, etc.) are tested because they provide certain benefits, such as operating over a wide pH range, avoiding or reducing premineralization or pH adjustment. And this, even if these organic coagulants are more expensive because they can bring several advantages in terms of performance. For colored to highly colored waters, iron-based coagulants are recommended: ferric chloride, ferric sulfate and ferrous sulfate, and with preference, the first of them. For flocculation, anionic polymers are recommended as flocculants to be tested first. The thickening of settling soil (sludge) and its conditioning are tested with cationic polymers instead.

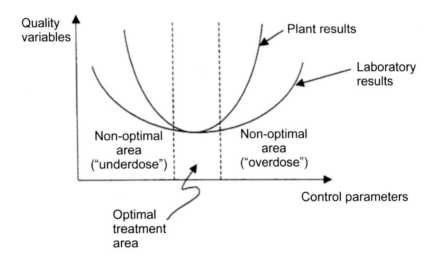

Figure 4.31. *Comparison of jar-test plant and laboratory results. For a color version of this figure, see www.iste.co.uk/gaid/watertreatment1.zip*

4.11.1. *The specific case of jar test for Actiflo®*

This method was developed to simulate the Actiflo® settling process. It is used for determining the coagulation pH, coagulant and polymer dosages to obtain the lowest turbidity and a high filterability index. In addition, the method can be used to monitor the performance of existing Actiflo® settlers and to diagnose operational problems. The proposed procedure takes into account Actiflo® hydraulic by respecting the contact times in the different basins in accordance with the design flow rate and respecting the injection order of chemical reagents and microsand. In a classic Actiflo®, operating at a flow rate corresponding to a superficial velocity of 40–60 mh^{-1} in the settling basin, the water residence times are:

– coagulation basin for 2 min;

– injection basin for 2 min;

– maturation basin 6 min;

– settling basin for 3–5 min.

Some variants exist with a common injection and maturation tank whose time varies between 6 and 8 min. In cold waters, it is often necessary to allow for a longer contact time between the coagulant and the raw water so that the flocs have enough time to form. To simulate a coagulant injection at a point upstream of the coagulation basin, additional mixing time can be added before the jar-test rapid mixing. In this way, it may be possible to determine the required contact time between raw water and the coagulant to obtain better coagulation and to infer the location of the injection point to be used in the plant. The new or used microsand (at least 10 times) is properly rinsed to eliminate the turbidity caused by sand micrograins or by the particles that may remain in suspension after the jar tests. For a volume of sand of 20 mL, we need to wash 10 times with 100 mL of distilled water. One may also take microsand directly from the underflow of the plant's hydrocyclones, and we need to be careful to properly clean it. A polymer stock solution was prepared at a 1–2 g·L^{-1} concentration. The solution was carefully mixed using the magnetic stirrer for 1–2 h until the polymer is well dissolved. Then, 100 mg·L^{-1} of polymer solution was prepared from demineralized water or the plant's service water. To adjust the coagulation pH (usually between 6 and 7), it is necessary to carry out preliminary tests to determine the amount of chemicals required to adjust the coagulation pH. The coagulation pH is adjusted with an acid or a base. Extra support should be provided to be added under the beakers so that the mixer blades are less than 0.5 cm from the beakers' bottom. This operation will keep the microsand in suspension during the tests.

4.11.1.1. Laboratory procedure

1) Set the temperature of the thermostatic bath to the desired temperature.

2) Measure the decanted water monitoring parameters (turbidity, color, dissolved aluminum, etc.).

3) Fill 1 L beakers with raw water.

4) Arrange the beakers on the test bench.

5) Ensure that the temperature of the raw water contained in the beakers is at the desired temperature.

6) Position the mixers' paddles so that they are less than 0.5 cm from the bottom of the beakers. To do this, it may be necessary to raise the beakers.

7) Start the device and adjust the rotational velocity to 150 rpm^{-1}.

8) Add acid or a base if required to adjust the coagulation pH.

9) Add the coagulant simultaneously in all the beakers using 0, 0.5, 1, 2 micrometric syringes or 5 min after adding the acid and/or the base, depending on the time of the simulated conditions.

10) Two minutes after adding the coagulant, add the microsand quantity in the beakers (1–10 g) and 50% of the polymer dosage using a pipette. For example, for a polymer solution of 100 mg·L^{-1}, add 2 mL when the desired polymer dosage is 0.4 mg·L^{-1}.

11) Two minutes after adding the microsand, add the remaining polymer (50%) using a pipette.

12) Six minutes after adding the polymer, stop the mixture and allow to settle for 3 min.

13) Draw a sufficient sample of supernatant to perform the required measurements using a 100 mL syringe approximately 5–10 cm from the water surface. Take care to draw off the liquid slowly to avoid altering the water sample taken.

4.11.1.2. Analysis: turbidity, color and other parameters

1) Measure the turbidity (NTU) using the turbidimeter.

2) Measure the apparent color using the spectrometer.

3) To measure true color, the filtrate produced during the tests is used to determine the filterability index.

4) The pH is measured using a pH meter.

5) Other parameters that can be measured: total organic carbon, dissolved aluminum, and SSs.

4.11.1.3. Filterability index

1) Filter a 100 mL volume of deionized water through a 0.45 μm filter under vacuum at 635 mm Hg, or the vacuum the pump can provide. The time required to filter the sample is measured.

2) Filter a 100 mL volume of raw water through a 0.45 μm filter under vacuum at 635 mm Hg, or the vacuum the pump can provide. The time required to filter the sample is measured.

3) The 100 mL decanted water sample is filtered under vacuum at 635 mm Hg through a 0.45 μm filter. The time needed to filter the sample is measured.

The filterability index is the filtered water volume divided by the time required to filter the sample divided by the filter's effective filtration area. The index can nonetheless be expressed in seconds if the filter's area and the filtered volume are constant.

The filterability index reached using the 100 mL sample of demineralized water corresponds to the reference value the filterability index of decanted water should approach as closely as possible. The smaller the difference between the filterability indexes is, the easier it is to filter the settled water.

Table 4.10 presents a calculation mode for various coagulants depending on the dosages applied.

Al coagulant	Iron coagulant
$Al_2(SO_4)_3$, $14H_2O$ MM Al:27 g/mole MM $Al_2(SO_4)_3$:342 g/mole MM Al_2O_3:102 1 mg·L⁻¹Al $= 1 \times 102/54 = 1.9$ mg·L⁻¹ Al_2O_3 $Al_2(SO_4)_3$ dosage: $1.9 \times 342/102 = 6.3$ mg·L⁻¹ Al_2O_3 content:17% Solid technical product dosage: $1.9 \times 100/17 = 10.8$ mg·L⁻¹ Technical solution conc. ($14H_2O$): $342 + 252 = 594$ g·L⁻¹ $Al_2(SO_4)_3$ conc. in the com solution:594 × 6.3/10.8 = 346.5 g·L⁻¹ Al_2O_3 conc. in the com solution: $346.5 \times 102/342 = 103.3$ g·L⁻¹ Volume dosage:10.8/594 = 0.018 mL/L	$FeCl_3$ MM Fe: 56 g/mole MM $FeCl_3$: 162.5 g/mole 1 mg·L⁻¹Fe $= 1 \times 162.5/56 = 2.9$ mg·L⁻¹ pure $FeCl_3$ $FeCl_3$ dosage (41%): $2.9 \times 100/41 = 7.1$ mg·L⁻¹ $FeCl_3$ solution density:1,430 kg·m⁻³ Volume dosage: $7.1 \times 1,000/1,430 = 4.9$ mL/m³
$Al_2(SO_4)_3$ MM Al:27 g/mole MM $Al_2(SO_4)_3$:342 g/mole MM Al_2O_3:102 1 mg·L⁻¹Al $= 1 \times 102/54 = 1.9$ mg·L⁻¹ Al_2O_3 $Al_2(SO_4)_3$ dosage: $1.9 \times 342/102 = 6.3$ mg·L⁻¹ Al_2O_3 content: 8.5 % Solid technical product dosage: $1.9 \times 100/8.5 = 21.6$ mg·L⁻¹ Solution density: 1,330 kg·m⁻³ Volume dosage: 21.6/1330 = 0.016	$FeSO_4$,$7H_2O$ MM Fe: 56 g/mole MM $FeSO_4$: 152 g/mole MM $FeSO_4$,7 H_2O: 278 g/mole 1 mg·L⁻¹Fe $= 1 \times 152/56 = 2.7$ mg·L⁻¹ pure $FeSO_4$ $FeSO_4$ technical dosage: $2.7 \times 278/152 = 5$ mg·L⁻¹ Sol. density. $FeSO_4$, $7H_2O$: 900 kg·m⁻³ Volume dosage: $5 \times 1,000/900 = 5.55$ mL/m³ = 0.035 mL/L
PACL and equivalents MM Al: 27 g/mole MM Al_2O_3: 102 1 mg·L⁻¹Al $= 1 \times 102/54 = 1.9$ mg·L⁻¹ Al_2O_3 Al_2O_3 content: 9.5% Dose in coagulant technical product: $1.9 \times 100/9.5 = 19.9$ mg·L⁻¹ Solution density: 1,350 kg·m⁻³ Volume dosage: 19.9/1,350 = 0.015 mL/L	$Fe_2(SO_4)_3$ MM Fe: 56 g/mole MM $Fe_2(SO_4)_3$: 400 g/mole MM $Fe_2(SO_4)_3$,$7H_2O$: 526 g/mole 1 mg·L⁻¹ Fe $= 1 \times 400/112 = 3.57$ mg·L⁻¹ pure $Fe_2(SO_4)_3$ Technical dose $Fe_2(SO_4)_3$: $3.57 \times 526/112 = 16.77$ mg·L⁻¹ Sol. density. $Fe_2(SO_4)_3$: 1,500 kg·m⁻³ Volume dosage: $16.77 \times 1,000/1,500 = 11.1$ mL/m³ = 0.0115 mL/L

Table 4.10. *Calculation of coagulant dosage*

4.12. References

Adachi, Y., Cohen Stuart, M.A., Fokkink, R. (1994). Dynamic aspects of bridging flocculation studied using standardized mixing. *Journal of Colloid and Interface Science*, 167, 346–351.

Amirtharajah, A. and O'Melia, C.R. (1990). Coagulation processes: Destabilization, mixing, and flocculation. In *Water Quality & Treatment*, 4th edition, Pontius, F.W. (ed.), McGraw-Hill, New York.

Bache, D.H. (2004). Floc rupture and turbulence: A framework for analysis. *Chem. Eng. Sci.*, 59, 2521–2534.

Baghvand, A., Daryabeigi Zand, A., Mehrdadi, N., Karbassi, A. (2010). Optimizing coagulation process for low to high turbidity waters using aluminum and iron salts. *American Journal of Environmental Sciences*, 6(5), 442–448.

Bosisio, M. and Milette, G.A. (1989). Études comparatives des performances de deux coagulants : le sulfate d'aluminium (alun) et le polyaluminium-silicate-sulfate (PASS). *Sciences et techniques de l'eau*, 22(4), 325–331.

Bottero, J-Y. and Lartiges, B. (1993). Séparation liquide/solide par coagulation-floculation : les coagulants-floculants, mécanismes d'agrégation, structure et densité des flocs. *Sciences Géologiques, bulletins et mémoires*, 46(1/4), 163–174.

Bratby, J. (2006). *Coagulation and Flocculation in Water and Wastewater Treatment*. IWA Publishing, London and Seattle.

Bridgeman, J., Jefferson, B., Parsons, S.A. (2010). The development and application of CFD models for water treatment flocculators. *Adv. Eng. Software*, 41, 99–109.

Brejchova, D. and Wiesner, M.R. (1982). Effect of delaying the addition of polymeric coagulant-aid on settled water turbidity. *Water Science Technology*, 26(9–II), 2281–2284.

Camp, T.R. and Stein, P.C. (1953). Velocity gradient and internal work in fluid motion. *Journal of Boston Society of Civil Engineering*, 10(30), 219–223.

Camp, T.R., Rost, D.K., Bhosta, B.V. (1940). Effects of temperature on the rate of floc formation. *J. Am. Water Works Assoc.*, 32(5), 913–927.

Cleasby, J.L. (1984). Is velocity gradient a valid turbulent flocculation parameter? *Journal of Environmental Engineering*, 10(5), 875–897.

Courcier, J-P., Faivre, M., Gaid, K., Holtz, C., Papet, J.-F. (2001). Coagulation, floculation, décantation et flottation. Technical document 43, Compagnie générale des eaux.

Crittenden, J.C., Trussell, R.R., Hand, D.W. (2012). *Water Treatment: Principles and Design*, 3rd edition. John Wiley & Sons, Hoboken.

David, W. (2010). *Hendricks Fundamentals of Water Treatment Unit Processes: Physical, Chemical, and Biological*. IWA Publishing, London and Seattle.

De Dianous, F. and Dernaucourt, J.C. (1991). Advantages of weighted flocculation in water treatment. *Water Supply*, 9, 543–546.

Dihang, M.D. (2007). Mécanismes de coagulation et de floculation de suspensions d'argiles diluées rencontrées en traitement des eaux. PhD Thesis, Université Paul Sabatier, Toulouse.

Dlamini, S.P., Harhoff, J., Mamba, B.B., Van Staden, S. (2013). The response of typical South African raw waters to enhanced coagulation. *Water Science & Technology*, 13(1), 20–27.

Ebeling, J.M., Sibrell, P.L., Ogden, S.R., Summerfelt, S.T. (2003). Evaluation of chemical coagulation & flocculation aids for the removal of suspended solids and phosphorus from intensive recirculating aquaculture effluent discharge. *Aquacultural Engineering*, 29, 23–42.

Edzwald, J.K. and Kaminski, G.S. (2009). A practical method to select coagulant. *Journal of NEWWA*, 11–27.

Gebbie, P. (2006). An operator's guide to water treatment coagulants. In *31st Annual Queensland Water Industry Workshop*, Central Queensland University, Rockhampton.

Gregory, J. (1989). Fundamentals of flocculation. *Crit. Rev. Environ. Control*, 19, 185–230.

Gregory, J. and Guibai, L. (1991). Effects of dosing and mixing conditions on polymer flocculation of concentrated suspensions. *Chem. Eng. Commun.*, 108, 3–21.

Haarhoff, J. and Cleasby, J.L. (1988). Comparing aluminum and iron coagulants for in-line filtration of cold-waters. *J. Am. Water Works Assoc.*, 80(4), 168–175.

Haarhoff, J. and Joubert, H. (1997). Determination of aggregation and breakup constants during flocculation. *Water Sci. Technol.*, 36, 33–40.

Hamaker, H. (1937). Attraction London – van der Waals between spherical particles. *Physica*, 4(10), 1058–1072.

Hanson, A.T. and Cleasby, I.L. (1990). The effects of temperature on turbulent flocculation: Fluid dynamics and chemistry. *J. Am. Water Works Assoc.*, 80(12), 56–73.

Jarvis, P., Jefferson, B., Parsons, S.A. (2005a). How the natural organic matter to coagulant ratio impacts on floc structural properties. *Environ. Sci. Technol.*, 39, 8919–8924.

Jarvis, P., Jefferson, B., Gregory, J., Parsons, S.A. (2005b). A review of floc strength and breakage. *Water Res.*, 39, 3121–3137.

Jarvis, P., Banks, J., Molinder, R., Stephenson, T., Parsons, S.A., (2008). Processes for enhanced NOM removal: Beyond Fe and Al coagulation. *Water Sci. Technol. Water Supply*, 8(6), 709–716.

Jekel, M.R. and Heinzmann, B. (1989). Residual aluminum in drinking-water treatment. *JWSRT-Aqua*, 38, 281–288.

Kaeding, U.W., Drikas, M., Dellaverde, P.J., Martin, D., Smith, M.K. (1992). A direct comparison between sulphate and polyaluminium chloride as coagulants in a water treatment plant. *Water Supply*, 10(4), 119–132.

Kimura, M., Matsui, Y., Kondo, K., Ishiokawa, T.B., Matsushita, T., Shirasaki, N. (2013). Minimizing residual aluminium concentration in treated water by tailoring properties of polyaluminium coagulants. *Water Res.*, 47, 2075–2084.

Lawler, F.D. (1993). Physical aspects of flocculation: From microscale to macroscale. *Water Res.*, 27, 165–180.

Lee, B.J. and Molz, F. (2014). Numerical simulation of turbulence-induced flocculation and sedimentation in a flocculant-aided sediment retention pond. *Env. Eng. Res.*, 19, 165–174.

Letterman, R.D., Amirtharajah, A., O'Meila, C.R. (2010). Coagulation and flocculation. In *Water Quality & Treatment: A Handbook on Drinking Water*, Edzwald., J. (ed.). McGraw-Hill, New York.

Lin, Q., Hanping, P., Huang, H., Liu, G., Yin, G. (2012). Flocculation mechanism by a novel combined aluminium-ferrous-starch flocculant (CAFS). *Water Science & Technoloy*, 65(12), 2169–2174.

Marques, R.O. and Ferreira Filho, S.S. (2017). Flocculation kinetics of low-turbidity raw water and the irreversible floc breakup process. *Environmental Technology*, 38(7), 901–910.

Narkis, N., Ghattas, B., Rebhün, M., Rubin, A.J. (1991). The mechanism of flocculation with alurninum salts in combination with polymeric flocculants as flocculant aids. *Water Supply*, 9, 37–44.

Oyegbile, B., Ay, P., Narra, S. (2016). Flocculation kinetics and hydrodynamic interactions in natural and engineered flow systems: A review. *Environ. Eng. Res.*, 21(1), 1–14.

Pernitsky, D.J. and Edzwald, S.K. (2006). Selection of alum and polyaluminium coagulants, principle and applications. *Journal of Water Supply, AQUA*, 55(2), 121–141.

Selvapathy, P. and Jayapal, R.M. (1992). Effect of polyelectrolytes on turbidity removed. *Water Supply*, 10(4), 175–178.

Smoluchowski, M. (1917). Modeling of coagulation equation. *Z. Phys. Chem.*, 92, 129–139.

Stumm, W. and Morgan, J.J. (1996). *Aquatic Chemistry: Chemical Equilibria and Rates in Natural Waters*. John Wiley & Sons, New York.

Tambo, N. (1991). Basic concepts and innovative of coagulation/flocculation. *Water Supply*, 9, 1–10.

Tambo, N. and Hozumi, H. (1979). Physical characteristics of flocs II. Strength of flocs. *Water Res.*, 13, 421–427.

Tambo, N. and Watanabe, Y. (1979). Physical characteristics of flocs I. The floc density functions and aluminium flocs. *Water Res.*, 13, 409–419.

Tardat-Henry, M. (1989). Évolution des dérivés de l'aluminium utilisés comme agents de coagulants. *Sciences et techniques de l'eau*, 22(4), 297–304.

Thomas, D.N., Judd, S.J., Fawcet, N. (1999). Flocculation modelling: A review. *Water Res.*, 33, 1579–1592.

Van Benschoten, S.E. and Edzwald, S.K. (1990a). Chemical aspects of coagulation using aluminium salts 1. Hydrolytic reactions of alum and polyaluminium chloride. *Water Res.*, 24(12), 1519–1526.

Van Benschoten, S.E. and Edzwald, S.K. (1990b). Chemical aspects of coagulation using aluminium salts II. Coagulation of fulvic acid using alum and polyaluminium chloride. *Water Res.*, 24(12), 1527–1535.

Van Benschoten, S.E., Edzwald, S.K., Rahman, M.A. (1992). Effects of temperature and pH on residual aluminium for alum and polyaluminium coagulants. *Water Supply*, 10(4), 49–54.

Vadasarukkai, Y.S., Gagnon, G.A., Campbell, D.R., Clark, S.C. (2011). Assessment of hydraulic flocculation processes using CFD. *J. Am. Water Works Assoc.*, 103(11), 66–80.

Wilkinson, K.J. and Reinhardt, A. (2005). Contrasting roles of natural organic matter on colloidal stabilization and flocculation. In *Flocculation in Natural and Engineered Environmental Systems*, Liss, S.N., Droppo, I.G., Leppard, G.G., Milligan, T.G. (eds). CRC Press, Boca Raton.

Wilson, L.D. (2014). An overview of coagulation-flocculation technology. *Water Conditioning & Purification*, Spotlight, April. 5.

Yan, M., Wang, D., Ni, J., Qu, J., Chow, C.W.K. (2008a). Relative importance of hydrolyzed Al (III) species during coagulation with polyaluminium chloride: A case study with typical micro-polluted source waters. *Colloid and Interface Science*, 316(2), 482–489.

Yan, M., Wang, D., Yu, J., Ni, J., Edwards, M., Qu, J. (2008b). Enhanced coagulation with polyaluminium chlorides: Role of Ph/alkalinity and speciation. *Chemosphere*, 71(9), 1665–1673.

Yukselen, M.A. and Gregory, J. (2004). The effect of rapid mixing on the break-up and re-formation of flocs. *J. Chem. Technol. Biotechnol.*, 79, 782–788.

Zhu, Z. (2014). Theory on orthokinetic flocculation of cohesive sediment: A review. *J. Geosci. Environ. Prot.*, 2, 13–23.

Algae

Hadjoudja, S., Deluchat, V., Baudu, M. (2010). Cell surface characterization of *Microcystis aeruginosa* and *Chlorella vulgaris*. *J. Colloid Interface Sci.*, 342, 293–299.

Henderson, R., Parsons, S.A., Jefferson, B. (2008). The impact of algal properties and pre-oxidation on solid-liquid separation of algae. *Water Res.*, 42, 1827–1845.

Henderson, R., Parsons, S.A., Jefferson, B. (2010). The impact of differing cell and allogenic organic matter (AOM) characteristics on the coagulation and flotation of algae. *Water Res.*, 44, 3617–3624.

Teixeira, M.R. and Rosa, M.J. (2007). Comparing dissolved air flotation and conventional sedimentation to remove cyanobacterial cells of Microcystis aeruginosa. Part II. The effect of water background organics. *Sep. Purif. Technol.*, 53, 126–134.

Teixeira, M.R., Sousa, V., Rosa, M.J. (2010). Investigating dissolved air flotation performance with cyanobacterial cells and filaments. *Water Res.*, 44, 337–3344.

5

Settling

In water treatment, settling is commonly used for removing particles that have been made decantable in the form of flocs by means of coagulation and flocculation. The settling stage must fulfill two main functions:

– separate the liquid–solid mixture;

– accumulate and thicken the mass of aggregated flocs to extract the solids in the smallest possible volumes and then send them to the corresponding sewage sludge treatment.

The settler's design, choice of equipment and shape must be designed in such a way that these two functions are respected.

5.1. The principles of settling

Floc removal from the liquid medium is relies solely on gravitational forces. The flocs likely to be sedimented correspond to various physical states. A close examination of a settling curve reveals the progressive transformation of particles within a settling tank. These transformations exclusively pertain to the physicochemical nature of flocs. Figure 5.1 illustrates the different aspects of settling.

There are four types of settling:

– the first corresponds to floc settling considered individually, depending on their density and size. During settling exclusively due to gravitational forces, particles are not hindered by other particles or by any external factors (such as the settling tank's wall). This is called free or grain settling;

– the second type corresponds to low-concentration solutions whose particles agglutinate or flocculate after collision. Their size increases as they fall, which also results in an increase in the settling rate. This is known as diffuse or coalescent settling;

– the third type relates to the global settling of a set of particles. In this case, each particle retains its position in relation to the others. Thus, a layer of particles is formed, which sediments at its own pace. This is called hindered settling;

– the fourth type is related to floc compression.

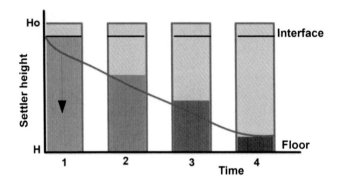

Figure 5.1. *Kynch-like construction. For a color version of this figure, see www.iste.co.uk/gaid/watertreatment1.zip*

The settling rate depends on the direction of liquid circulation (horizontally, vertically, in circular or spiral motion), the type of flow (laminar or turbulent), the medium's viscosity and the shape, diameter and density of the flocs obtained during flocculation.

A spherical particle with a density ρs falls increasingly quickly until the force of attraction and the force of resistance due to friction are balanced.

The falling velocity *vc* becomes constant and is equal to:

$$vc = \frac{g\,(\rho s - \rho)}{18\,\mu}\,d^2$$

where:

– *vc*: particle falling velocity (m·s^{-1});

– *g*: acceleration due to gravity (9.81 m·s^{-2});

– ρs and ρw: particle densities (flocs) and water (g·cm^{-3});

– μ: dynamic viscosity (1.1×10^{-3} Pa·s^{-1});

– d: particle diameter (m).

vc therefore depends on the particles' diameter (flocs), their density and water viscosity (itself dependent on temperature).

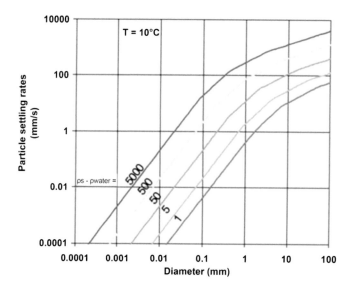

Figure 5.2. *Stokes velocities for particles with different diameters. For a color version of this figure, see www.iste.co.uk/gaid/watertreatment1.zip*

5.2. Horizontal settlers

5.2.1. *Principle*

Settling is based on Hazen's sediment load theory, according to which the smallest particle size to be removed is the one that can settle along its path. In a horizontal settler, for a length L and a water depth H, the smallest particle that will have time to settle is determined by the falling velocity (vc):

$$vc = v\frac{H}{L}$$

where v is the (horizontal) circulation velocity in the settler.

Thus, the design of a horizontal settler (also called a settling corridor) must encourage the flocs to enter the settler at a sufficient velocity to reach the tank's floor. The cross-sectional view of a basin (Figure 5.3) depicts the trajectories of the smallest particle and any particle, with their respective falling velocity.

Thus, for a floc to settle, it must reach the bottom of the settler before being evacuated, which results in the following ratio:

$$t = \frac{L}{v} \geq tc = \frac{H}{vc}$$

where:

– t: dwell time (h) in the settler;

– vc: floc or particle's falling velocity (m·s^{-1});

– H: total height (m);

– L: settler length L (m) (to outlet);

– v: flow rate (m·h^{-1});

– tc: floc falling time (h).

That is,

$$vc \geq \frac{H.v}{L} = \frac{Q.H}{L.H.l} = \frac{Q}{S}$$

In theory, the particles will only be retained in the settler if vc is $\geq v$

$$vc = \frac{v.H.l}{l.L} = v\frac{H}{L}$$

where:

– Q: treatment capacity (m^3·h^{-1});

– v: input velocity (m·h^{-1});

– H, l and L: settler height (m), width (m) and length (m);

– vc: floc falling velocity (m·s^{-1});

– S: settler's mirror surface (m^2).

The above equation means that the treatment capacity is proportional to the settler's mirror surface. The smallest flocs should settle, provided that their falling velocity is greater than or equal to Q/S. It is therefore important to control this parameter by obtaining the best possible flocculation, inducing flocs to have the largest possible diameter and leading to an increase in the Stokes drift velocity.

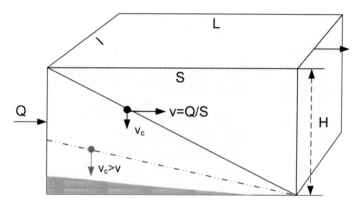

Figure 5.3. *Flocfall and circulation velocity in a horizontal settler. For a color version of this figure, see www.iste.co.uk/gaid/watertreatment1.zip*

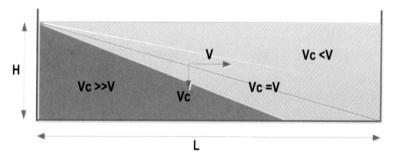

Figure 5.4. *Settling profile as a function of falling velocity and flow velocity. For a color version of this figure, see www.iste.co.uk/gaid/watertreatment1.zip*

In the case where the flocs' falling velocity is lower than the flow velocity (mirror settling velocity), flocs are found at the end of the settler. Turbidity can be degraded (Figure 5.4). When the flocs' falling velocity is greater than the flow velocity, the flocs settle very quickly at the beginning of the settling tank. This is the case for heavily loaded waters with suspended solids.

When the flocs' falling velocity is near or equal to the flow velocity, the flocs settle along the course of the water and are dispersed over two-thirds of the settler's length. This is the case for conventional waters, with variable turbidity between 20 and 100 NTU. Turbidity in settled water is <1 NTU.

A horizontal settler is considered to be divided into four zones:

– an inlet zone: the flow is uniformly distributed along a horizontal direction and assimilated to a piston flow;

– a settling zone: settling takes place, encouraging a liquid–solid separation that increases throughout the settling tank;

– an outlet zone: the flow converges toward the outlet spillways located on the settler's surface;

– a sewage sludge area: the settled matter accumulates along the water's path with a high concentration at the beginning of the settler. It is believed that once a floc reaches the settler's floor, it is effectively removed from the water.

Therefore, the settler's length takes this into account so that the smallest decantable particle can be removed from the water. The length can extend up to 80 m, whereas the maximum observed width is 20 m.

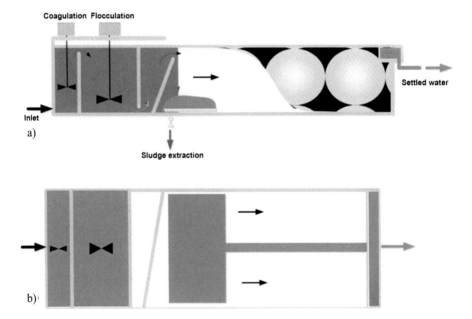

Figure 5.5. *Representation of a horizontal settler's operating principle: (a) side view and (b) view from above. For a color version of this figure, see www.iste.co.uk/gaid/watertreatment1.zip*

5.2.2. Design

One of the main parameters to consider is the mirror settling velocity $(m \cdot h^{-1})$ since it is related to the flocs' falling velocity. Indeed, the efficiency of settling is theoretically independent from the tank's depth, provided that the mirror settling velocity (flow velocity) is low enough for the settling flocs not to be resuspended by a turbulence effect. This is the reason why we take a preliminary approach, contemplating variable depths between 3.5 and 5 m.

The mirror settling velocity $(m \cdot h^{-1})$ is given by:

$$v = \frac{Q}{S}$$

The retention time (h) is given by:

$$t = \frac{S.H}{Q}$$

where:

– Q: inlet flow $(m^3 \cdot h^{-1})$;

– S: mirror surface (m^2);

– H: water depth (m).

Retention times in the horizontal settlers vary between 2 and 4 h. Some settlers even have retention times longer than 12 h.

The input velocity $(m \cdot h^{-1})$ must be correctly set to prevent excessive dimensioning from causing settled flocs to be resuspended. This stipulation influences the choice of length/width and height/width ratios for this type of settler. It is based on the ratio as follows:

$$v\max = \frac{Q}{H.l} = \frac{L}{t}$$

Here, l is the settler's width (m). In practice, the L/l ratio varies between 3 and 5, while 4 is the recommended value.

The performances of horizontal settlers are sensitive to temperature differences due to the slow flow velocity $(<0.5 \ m \cdot h^{-1})$, which characterizes these structures, particularly during warm periods. A temperature difference of only 1°C is sometimes enough to cause a drop in efficiency due to the changes in density.

Parameters	Dimensioning
Mirror settling velocity ($m \cdot h^{-1}$)	0.5–1.2
Width (m)	5–20
Length (m)	20–80
Water level (m)	3–5
Retention time (h)	2–4
L/H ratio	<18
L/l ratio	3–5

Table 5.1. *A few dimensioning parameters for horizontal settlers*

5.2.3. *Implementation*

It is clear that ideal settling is impossible to achieve at the treatment plant due to flow rate variations, strong variations in inlet turbidities, the difficulty in obtaining a homogeneously distributed flow, avoiding short circuits caused by sewage sludge deposits, etc.

Nevertheless, the manufacturer implements all the technical means to obtain the best liquid/solid separation. The settling tank's design should be such that the smallest flocs can be retained based on contact times, which will ensure high floc removal. A well-sized horizontal settler whose operating conditions for coagulation and flocculation (with a polymer) are optimal has an outlet turbidity <0.4 NTU for a 100 NTU inlet, and an outlet turbidity <1 NTU for a 500 NTU inlet.

The optimum mirror settling velocities are lower than 0.5 $m \cdot h^{-1}$. When this velocity rises to 1 $m \cdot h^{-1}$, the efficiency decreases and reaches approximately 98.5–99% for a 500 $mg \cdot L^{-1}$ load.

The equipment for feeding settlers must include submerged inlets to enable a stable and uniform flow. Settled water is collected in perforated pipes located at the end of the settlers or in troughs covering approximately 10–20% of the surface.

There are two types of horizontal settlers with scraper mechanisms: scraper bridge and chain scraper settlers.

Scraper bridges move to-and-fro and perform scraping following countercurrent motion. During the scraping period, the velocity cannot exceed 3 $cm \cdot s^{-1}$, which limits scraping devices to 60 m.

a)   b)

Figure 5.6. *Corridor settlers operated by Veolia some years ago:
(a) Shanghai (China) and (b) Arcuda (Romania). For a color version
of this figure, see www.iste.co.uk/gaid/watertreatment1.zip*

Chain settlers enable a continuous scraping of sludge and floating bodies due to a series of scrapers mounted between two endless parallel chains, which rotate along the basin's vertical walls.

There are also trac-vac-type devices that suck up the floc deposited all along the settling tank's floor using a suction pump that circulates horizontally along the structure.

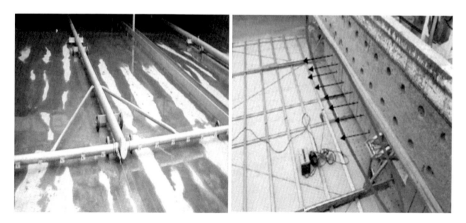

Figure 5.7. *Suction equipment for deposited flocs (trac-vac type). For a color
version of this figure, see www.iste.co.uk/gaid/watertreatment1.zip*

Veolia has built and operates horizontal settlers (also called corridor settlers) not only in France but also in many other countries.

This type of settler is interesting when the ground surface is available. Since this is becoming less frequent, it is advisable to envision the design of more compact settling tanks, which can offer an equivalent performance.

Horizontal settlers may have a reduced performance if the laminar flow is not evenly distributed, thus creating hydraulic short circuits that can disrupt their proper operation. In addition, sludge extraction must be well developed along the entire floor surface to avoid accumulation in various settler areas, which could cause floc leaks. Turbidity at the outlet could then be degraded. Finally, the concentration of the sewage sludge extracted is relatively low, which implies a larger dimensioning of the thickening structure.

For all these reasons, the production of circular and lamella settlers has developed, together with the marketing of new compact and high-performance technologies.

5.3. Lamella settlers

There are two types of conventional lamella settlers, which are differentiated depending on the water flow's direction.

When the water flows (downward) in the same direction as flocs do, settlers are cocurrent. When the water flows in the opposite direction to the flocs, settlers are countercurrent.

– Although cocurrent systems facilitate floc removal, short circuits are often observed.

– In countercurrent systems, the admission of flocculated water is carried out below the lamella block. This results in a crossing between the sewage sludge that slides along the plates and migrates to the settler's bottom. The water is evacuated by the top and then collected in troughs. The clarified water is obtained in the area where floc concentration is the lowest, which results in better clarified water. In addition, operating in countercurrent involves a simpler design than the co-current mode. Hydraulic flow control upward is easier than the downward flow control present in co-current systems. This ease compensates for the difference in the projected surface of these two systems.

This is one of the reasons why Veolia opted for this type of technology.

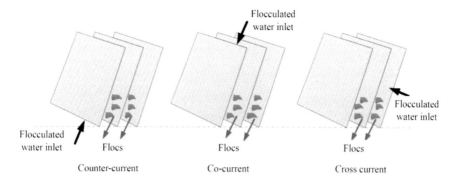

Figure 5.8. *Different feeding systems for lamella blocks. For a color version of this figure, see www.iste.co.uk/gaid/watertreatment1.zip*

There are also cross-current lamella settlers where the floc flow is perpendicular to the water's direction. The countercurrent design is often preferred due to its simplicity.

Veolia lamella settlers fall into several categories:

– with or without a sewage sludge bed;

– with or without weighted flocs.

5.3.1. *Theory and principle*

If the same theory used for horizontal settlers is applied to lamella settlers, the parameter relating to the presence of lamellae must be brought into play.

A lamella module is a piece of equipment made up of plates with an inclined surface (either flat or corrugated) and piled up within a short distance from one another. A lamella settler can be described as the compilation of a large number of elementary settling tanks, each of which comprises the volume between two settling plates.

Thus, two back-to-back plates constitute a module composed of hexagonal sections. The flocculated water is distributed under the modules through an orifice that has the particularity of extending across the entire device's width. Water flows upward (countercurrent motion) between the plates in the opposite direction to the flocs, which settle on the plates and slide downward due to the effect of gravity.

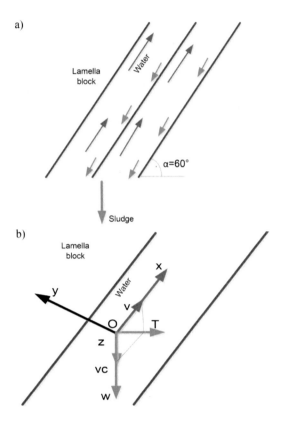

Figure 5.9. *Countercurrent operating principle (a) and spatial representation of liquid–solid separation (b). For a color version of this figure, see www.iste.co.uk/gaid/watertreatment1.zip*

The space formed between each lamella plate represents a modular settler whose Hazen velocity (associated with the cutoff threshold) is characterized by the ratio between the inflow and the lamella plate's projected surface. It is expressed by the ratio

$$v \ (\text{m} \cdot \text{h}^{-1}) = Q/Sl \cos \alpha$$

where:

– Q: flow $(\text{m}^3 \cdot \text{h}^{-1})$;

– Sl: lamella plates' surface (m^2);

– α: angle of inclination of the lamella plates in relation to the horizontal plane.

The settler's depth has no influence on the switching power, provided that the deposited material is evacuated at the same pace it accumulates and does not disturb the inlet water flow.

The combined use of a high mirror velocity associated with inclined plates (60°) enables continuous self-scraping. There is little floc accumulation on the plates. If we define a coordinate system Ox, Oy, Oz, Ow, then:

– Ox is parallel to the lamella plates;

– Oy is perpendicular to the lamella plates;

– Oz is perpendicular to the xOy plate;

– Ow is perpendicular to xOy and is related to the flow direction of the deposited floc on the plate.

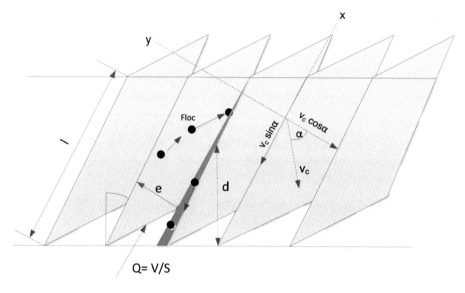

Figure 5.10. *Floc trajectory on lamella plates. For a color version of this figure, see www.iste.co.uk/gaid/watertreatment1.zip*

The water moves along Ox (clarified water outlet) with a flow velocity between two plates (Figure 5.10), and flocs are subjected to a Hazen velocity (vertically along Ow), corresponding to the direction and fall of the deposited floc. The angle α is determined so that the deposited floc can slide to the bottom without accumulating at a specific location on the plates. This is why the total projected area is calculated in such a way that the laminar regime (Reynolds numbers <500 and Froude numbers $>10^{-5}$) enables gravitational forces to drag the flocs down the settling tank.

Figure 5.10 illustrates the behavior of flocs between the two plates. α is the plates' (or lamellae) angle of inclination, v is the mirror settling velocity ($m \cdot h^{-1}$), S is the settler area (m^2), Q is the cross flow ($m^3 \cdot h^{-1}$), e is the gap between plates (m), d is the vertical distance between two plates (m), vc is the flocs' falling velocity (mh^{-1}), H is the plate's straight length (m) and l is the plate's projected length (m).

When a particle moves along the lamella plates, the time required for the particle to move through the lamellae is $L/(v - vc\ \sin\alpha)$. The time required to settle on the lower lamella is $H/vc\ \cos\alpha$.

Thus, the smallest particle that can be removed by the lamella settler is determined by the falling velocity:

$$vc = v\frac{H}{L\cos\alpha + H\sin\alpha}$$

If $\alpha = 0$, we obtain $vc = vH/L$ related to the horizontal settling tanks.

If $\alpha = 90°$, then $vc = v$. In this case, the flocs' falling velocity must be greater than the water flow velocity through the lamella plates for settling to take place. It is only if $vc < v$ that the particles are removed.

5.3.2. *Basic design for lamella settlers*

Particle removal (flocs, suspended solids) in lamella settlers depends on the developed surface but not on the settlers' depth (Hazen, Camp). This can be observed, for example, when tilting a tube containing flocs compared to the same tube in a vertical position.

5.3.2.1. *The case of honeycombed lamella plates*

For each type of lamella, there is a specific ratio between the total projected area (TPA) and the mirror surface (MS), which is key to the efficiency of floc separation. In this way, we can determine this ratio for each type of lamella and use this value for calculating velocity values.

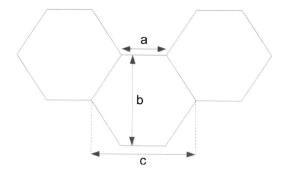

Figure 5.11. *Characteristics of honeycombed lamellae*

We have:

– Lp: lamella plate length (mm);

– a: settling width in a honeycomb cell (mm);

– b: maximum orthogonal depth (mm);

– c: maximum settling width (mm);

– d: average width in a honeycomb cell (mm): (c + a)/2;

– circulation surface (mm²): b·d;

– mirror surface of a honeycomb cell (mirror s; mm²): circulation surface/sinα;

– settling area (mm²): L × c;

– total projected area (TPA, mm²): settling area × cosα.

The TPA/mirror surface ratio is deduced there from.

It is equivalent to the ratio between the mirror velocity (v, m·h⁻¹) and Hazen velocity (vH, m·h⁻¹)

TPA/mirror surface = v/vH

$$\frac{TPA}{MS} = \frac{v}{vH}$$

Table 5.2 shows an example of dimensioning parameters based on the characteristics of Greca-type hexagonal lamella plates.

Parameters	Values	
Angle of inclination	60	60
Settling width in a honeycomb cell (mm)	25	25
Maximum orthogonal depth (mm)	36.2	36.2
Maximum settling width (mm)	45	45
Lamella plate length (mm)	1,000	1,400
Lamella plate height (mm)	860	1,204
Mean width in a honeycomb cell (mm)	35	35
Circulation surface (mm²)	1,267	1,267
Mirror surface of a honeycomb cell (with MS, mm²)	1,463	1,463
Settling area (mm²)	45,000	63,000
Total projected area (TPA, mm²)	22,500	31,500
TPA/mirror surface ratio	15.38	21.53
Hazen velocity (m·h⁻¹)	0.8	0.8
Mirror velocity (m·h⁻¹)	12.30	17.2
Wet perimeter (m)	133	133
Hydraulic diameter (m)	0.038	0.038
Reynolds (Re)	131	183
Froude (Fr)	0.006	0.009

Table 5.2. *Design example based on the characteristics of lamella plates*

For each type of hexagonal lamella for which the parameters a, b and c are known or can be measured, it is easy to determine this TPA/MS ratio, which gives access to the Hazen velocity and/or the mirror velocity. Thus, for a Hazen velocity of 0.8 m·h⁻¹, the mirror velocity is 12.28 m·h⁻¹ with a TPA/MS ratio equal to 15.36. This calculated ratio is associated with the lamella plates' length, following the (TPA × Lp)/MS formulae.

An increase in this length implies an increase in the TPA/MS ratio. For a given Hazen velocity, the mirror settling velocity can be increased. For a gap between two identical plates, an increase in length involves a longer trajectory for the flocs and a higher cutoff threshold.

For example, for a length of 1.40 m, the TPA/MS ratio rises from 15.36 to 21.50. For a Hazen velocity equal to 0.8 m·h⁻¹, the mirror velocity increases to 17.2 m·h⁻¹. This implies a more compact settler to obtain equivalent performance.

Keeping the mirror velocity at 12.30 m·h⁻¹, increasing the length to 1.40 m leads to a higher total projected area and therefore to a Hazen velocity that drops to 0.57 m·h⁻¹. Performance is thus significantly improved.

Veolia has built and operates many lamella settlers (Multiflo®) based on a 1.40 m plate length.

5.3.2.2. The case of flat lamella plates

Particle removal (flocs, suspended solids) in lamella settlers fitted with flat plates is presented in Table 5.3.

Parameters	Equations
Hazen velocity (m·h)	vH = Q/Sp
Projected area of a plate (m²)	Sp = Lp.lp. cosα
Total projected area (m²)	Sp = n.Lp.lp. cosα
Mirror velocity	V = Q/S
Angle of inclination	α = 60°
Projected length of a plate	L = H/sinα

Table 5.3. *Dimensioning parameters of flat lamella plates*

In the above table, Lp is the plate length (m), lp is the plate width (m) and n is the number of plates.

Thus, the design of lamella settlers takes two parameters into consideration. These are the Hazen velocity $(m·h)^{-1}$ and the mirror velocity $(m·h^{-1})$.

The mirror velocity is given by:

$$v\ (m.h-1) = \frac{Q}{MS}$$

The Hazen velocity $(m·h^{-1})$ is given by:

$$vH\ (m.h-1) = \frac{Q}{TPA} = \frac{Q}{nSp.\cos\alpha}$$

where:

– MS: mirror surface (m²);

– Sp: plate surface (length × width in m²);

– TPA: total projected area (m²);

– e: gap between two plates (m);

– L: settler length (m);

– n: number of plates;

– Q: inlet flow (m³·h⁻¹).

It is possible to calculate vH using $(n - 1)$ plates so as not to count the one which is at the end of the settling tank, which is ineffective. However, the overall calculation remains equivalent, especially for treatment plants working with large flows.

The number of plates n is equal to

$$n = \frac{Lsin\alpha}{e}$$

An accurate calculation of L involves only counting the effective length containing lamella plates. Due to their inclination, a small area at the end of the settler is not covered.

For honeycombed hexagonal lamella, there is a specific ratio between the total projected area (TPA)/mirror surface, which is involved in floc separation efficiency.

We have:

– Lp: lamella plate length (mm);

– lp: lamella plate width (mm);

– e: gap between two plates (mm);

– circulation surface (mm²): e.lp;

– mirror surface of a plate between two plates e.lp/sinα;

– settling area (mm²): Lp.lp;

– total projected area (TPA, mm²): settling area × cosα.

The TPA/mirror surface ratio is deduced there from.

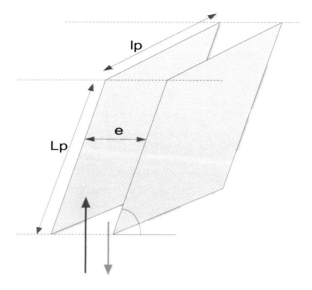

Figure 5.12. *Characteristics of flat lamella plates*

It is equivalent to the ratio between the mirror velocity (vm, m·h^{-1}) and Hazen velocity (VH, m·h^{-1})

TPA/MS = v/vH

$$\frac{TPA}{MS} = \frac{v}{vH}$$

Table 5.4 shows an example of dimensioning parameters based on the characteristics of Greca-type hexagonal lamella plates.

With a plate length of 2.5 m, the Re number is <500 and the hydraulic regime remains in the laminar regime. It is therefore possible to increase the Hazen velocity and the mirror velocity, provided that the values of Re and Fr are respected.

The gap between two flat plates is generally between 3 and 5 cm, with an angle of inclination of 60°. The plate length varies between 2 and 2.5 m. The total projected area is the plate's area multiplied by the number of plates and by cosα. The gap is in the range of 5 cm.

Parameters	Values	
Angle of inclination	60	60
Lamella plate width (mm)	1,030	1,030
Lamella plate length (mm)	2,000	2,500
Gap between two plates (mm)	50	50
Circulation surface (mm²)	51,500	51,500
Mirror surface between two plates (mirror s, mm²)	59,467	59,647
Settling area (mm²)	2,060,000	2,575,000
Total projected area (TPA, mm²)	1,030,000	1,287,500
Circulation surface/mirror surface	0.866	0.866
TPA/mirror surface ratio	17.32	21.651
Hazen velocity ($m \cdot h^{-1}$)	0.4	0.4
Mirror velocity ($m \cdot h^{-1}$)	6.93	8.66
Flow velocity ($m \cdot h^{-1}$)	0.00222	0.00278
Wet perimeter (m)	2,160	2,160
Hydraulic diameter (m)	0.095	0.095
Kinematic viscosity ($m^2 \cdot s^{-1}$)	1.15×10^{-6}	1.15×10^{-6}
Reynolds	184	230
Froude	0.002	0.003

Table 5.4. *Design example based on the characteristics of lamella plates*

5.3.3. *Implementation*

The retention time in lamella settlers is considerably reduced (less than 1 h, or even less than 15 min, see Actiflo sheet) when compared to horizontal settling tanks. The implementation of lamellae remarkably reduces the Reynolds number, thus limiting turbulence, which has a direct impact on residual turbidity.

The small size of lamella settling systems makes them easy to fit into buildings, thus facilitating the work of operators. Not only do they prevent the development of algae on the structures' walls, but they are also less sensitive to bad weather (black ice, ice).

Figure 5.13. *Transformation by Veolia of a horizontal settler into a lamella settler at Punta Lara plant (Argentina). For a color version of this figure, see www.iste.co.uk/gaid/watertreatment1.zip*

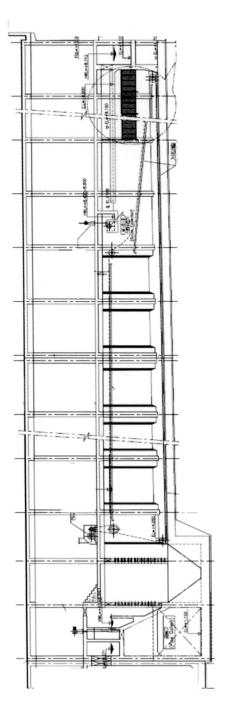

Figure 5.14. *Transformation by Veolia of a horizontal settler into a lamella settler at a Gulpocheon plant (Chile)*

It is important for lamella plates to constitute a uniform lamella block. If the plates are not sufficiently contiguous to one another, the upward flow of water can put the flocs back in suspension and degrade the quality of the settled water (turbidity).

5.4. Veolia technologies

5.4.1. *Lamella settlers: Multiflo® settler*

The Multiflo® settler combines the coagulation, flocculation and countercurrent lamella settling stages in a single unit. It thus brings together two areas under one piece of work: the areas where the physicochemical reactions take place (coagulation–flocculation) and the physical separation of flocs (lamella settling).

The Multiflo® settler applied to drinking water can be divided into two types:

– the Multiflo® Duo with turbomix (optional), tremies or harrow scraper;

– Multiflo® Trio with sewage sludge bed, sludge recirculation and turbomix.

In both cases, the flocculated water is distributed under the modules through an orifice that has the particularity of extending across the entire device's width (flow baffle).

5.4.1.1. *The Multiflo® Duo settler*

The Multiflo® Duo lamella settler ensures three main functions:

– occupies a minimum of space by exploiting the highest possible settling rate;

– retains as many particles in suspension as possible;

– concentrates the collected sewage sludge.

The settling zone is equipped with hexagonal modules made up of inclined lamella plates (60°). Thus, two side-by-side lamella plates form a module made up of ducts with a hexagonal section. At the base of the lamella plates, the floc deposition mechanism takes place continuously, despite the upstream water current. This is explained by the fact that, in this zone, the falling floc is made up of large clusters that have formed by accumulation, sliding along the plates. The rising velocity of water does not hinder settling.

To avoid floc shearing during the interaction of water with floc currents, the plates of Multiflo® Duo are sufficiently close together (<45 mm). This small spacing improves the cutoff threshold while avoiding floc shear phenomena when they slide toward the settling tank's bottom.

The water circulates upstream between the lamella plates in the opposite direction (countercurrent) to the flocs, which settle on the plates and slide downward due to the effect of gravity. Flocs arrive under the lamella plates, whose total surface is calculated in such a way that the laminar regime (Reynolds number <200) in vigor invites gravitational forces to cause the flocs deposited on the lamella plates to slide.

The juxtaposition of inclined and aerodynamic lamella plates creates independent modules, thus constituting numerous elementary settling units. The lamella surface also improves the hydraulic distribution and reduces the effects of turbulence, thus facilitating the settling of less decantable flocs.

Lamellar plates provide a large settling surface, corresponding to the projected surface of all the lamellae, compared with the conventional settlers where only the mirror surface is taken account. This results in a ground occupation 10–20 times smaller than with a non-lamella settling unit.

The reduced size of the mirror surface and the sewage sludge accumulation zone, as well as the distribution of the internal flows, ensure homogeneous and regular operation. The settling area's feed zone covers its entire width. In the longitudinal direction, it spans between the flooded spillway and the wall delimiting the lamella zone.

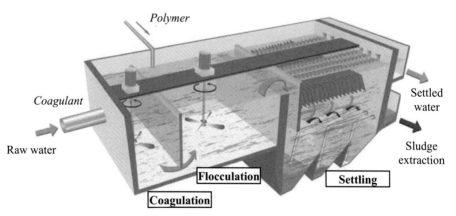

Figure 5.15. Multiflo® Duo settler with hoppers. For a color version of this figure, see www.iste.co.uk/gaid/watertreatment1.zip

In the Multiflo® Duo settling tank, three compromises are implemented:

– an optimal lamellae height (1 m or 1.40 m), being careful not to resuspend the deposited floc;

– the evacuation of the floc deposited by the gravitational flow and the use of an optimum angle of inclination;

– a quick reaction to raw water variability to consistently produce high-quality settled water.

The collection of settled water can be performed by perforated pipes or overflowing troughs fitted with trapezoidal or rectangular notches. The parameters of the design are summarized in Table 5.5.

Parameters	Multiflo® Duo	
	Clarification	Clarification
Use of polymer	No	Yes
Maximum mirror velocity	8 m·h^{-1}	12–18 m·h^{-1}
Sludge recovery	<= 2 × 3 hoppers Trac-vac Harrow circular scraper	<= 2 × 3 hoppers Trac-vac Harrow circular scraper
Conc. sludge	5–20 g·L^{-1}	5–20 g·L^{-1}
Water loss	3%	3%
Flocculation time (depending on T °C)	10 – 30 min	10 – 30 min
Lamellae gap	30–45 mm	30–45 mm
Lamellae length	1–1.5 m	1–1.5 m
TPA/MS	>15 m^2/m^2	>15 m^2/m^2
Expected performance (raw water → settled water)	100 → 3 NTU	200 → < 3 NTU 1,000 → 5 NTU

Table 5.5. *Design parameters for Multiflo® Duo settlers*

5.4.1.1.1. Performance of the Multiflo® Duo settler

The treatment results on heavily loaded water (350 NTU, Sungai Selangor River, Malaysia) show that the Multiflo® Duo settler releases clarified water with a turbidity between 2.5 and 3 NTU, operating with a Hazen velocity of 1.25 m·h^{-1} and a mirror velocity equal to 20 m·h^{-1}. Figure 5.16 clearly shows that when working with a relatively high Hazen velocity, the quality of the treated water always complies with the guarantees requested. On the Johor River site (Malaysia), the calculated Hazen velocity is 0.91 m·h^{-1} (mirror velocity = 15.26 m·h^{-1}) for the maximum flow of 475,000 m^3·j^{-1}.

The Multiflo® Duo settler is specifically adapted for the treatment of water with medium or high turbidity (50–4,000 mg·L^{-1} suspended solids) and produces water with < 1–3 NTU turbidity, depending on the quality of raw water.

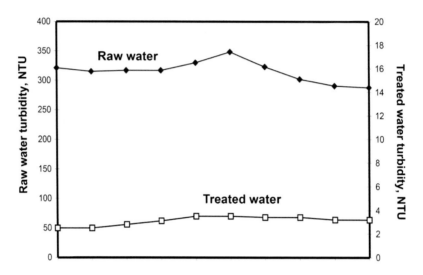

Figure 5.16. *Turbidity removal (Malaysia) with a Multiflo® Duo. For a color version of this figure, see www.iste.co.uk/gaid/watertreatment1.zip*

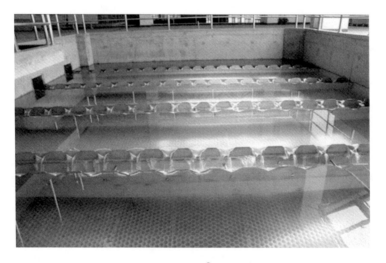

Figure 5.17. *Mirror surface of a Multiflo® Duo settler (France). For a color version of this figure, see www.iste.co.uk/gaid/watertreatment1.zip*

Dimensioning example of a Multiflo® Duo settler: inlet flow: 500 m³·h⁻¹, T °C: 12°C.

Coagulation	
Parameters	**Values**
Coagulator unit	1
Reaction time (min)	2
Area (m²)	2.82
Volume (m³)	16.9
Water level (m)	6
$G = ((P/\mu \cdot V)1/2)$ (s⁻¹)	283
Impeller power (W)	1,358
μ dynamic viscosity (kg·m⁻¹·s⁻¹)	0.001
Np power number	0.7
Impeller velocity (min⁻¹)	167.7
ρ water (kg·m⁻³)	1,000
Paddle diameter (m)	0.6
Gt	34,531
Flocculation	
Parameters	**Values**
Flocculator unit	1
Reaction time (min)	30
Area (m²)	41
Volume (m³)	246
Water level (m)	6
$G = ((P/\mu \cdot V)1/2)$ (s⁻¹)	50.3
Impeller power (W)	250
μ dynamic viscosity (kg·m⁻¹·s⁻¹)	0.001
Np power number	1.0
Impeller velocity (min⁻¹)	23.4
ρ water (kg·m⁻³)	1,000
Paddle diameter (m)	1.6
Gt	89,019

Settling	
Parameters	**Values**
Settler unit	1
Mirror surface (m²)	32.2
Unit length (m)	9.20
Non-effective length (m)	0.5
Total length (m)	9.70
Lamella length (m)	1.00
Gap (mm)	36
TPA/Ms	15.38
Total projected area (m²)	495.2
Hazen velocity (m·h^{-1})	0.99
Mirror velocity (m·h^{-1})	15.5
Retention time (min)	55.3
Settling volume (m³)	460.8

Table 5.6. *Design parameters for the Multiflo® Duo settler*

Figure 5.18. *Photo of the Chengdu plant (China 420,000 m³·Γ^{-1}) with 10 Multiflo® Duo settlers. For a color version of this figure, see www.iste.co.uk/gaid/watertreatment1.zip*

Figure 5.19. *Photo of a Multiflo® settler in Turkey. For a color version of this figure, see www.iste.co.uk/gaid/watertreatment1.zip*

5.4.1.2. *The Multiflo® Trio settler*

The Multiflo® Trio settler is a settler-thickener with the same basic characteristics as the Multiflo® Duo settler. This includes coagulation, flocculation and settling. The water circulates from bottom to top between the lamella plates in the opposite direction (countercurrent) of the flocs, which settle on the plates and slide downward under the effect of gravity.

Flocs arrive under the lamella plates, whose total surface is calculated in such a way that the laminar regime (Reynolds number <200) in vigor invites gravitational forces to cause the flocs deposited on the lamella plates to slide.

The Multiflo® Trio settler is characterized by the presence of a fluidized sludge bed at the level of the flocculator, which accelerates flocculation during its passage through raw water, producing denser and more easily separable flocs. By definition, the flocculation stage is intended to form large aggregates to propitiate rapid settling.

In periods of low turbidity (<10 NTU), the formation of these aggregates is not easy because there are no sufficient flocs to encourage the meeting and collisions between flocs. This was described when discussing orthokinetic and perikinetic flocculation mechanisms.

As such, sludge recirculation at the level of the flocculation tank increases the number of particles and accelerates collisions. It improves the treatment's effectiveness:

– during periods of low turbidity;

– in cold periods <5°C;

– during significant variations in the quality of raw water.

Sewage sludge recirculation (0–10%) increases the concentration in the flocculation tank and, consequently, the final concentration for the same sludge retention time. This produces densified flocs that increase the sedimentation rate. In periods of low turbidity, it prevents reagent overdosing on the part of operators who tend to abuse the use of products to increase floc volume and ensure better settling.

The efficiency of the Multiflo® Trio settler is also based on optimized flocculation, which includes specific equipment such as the Turbomix mixer. The flocculation reaction is undoubtedly effective when the coagulant and the polymer fully react to destabilize colloidal particles. Turbomix (Figure 5.20) makes it possible to promote homogeneity in the flocculation tank by ensuring the polymer's distribution to avoid by-passes and dead zones in this tank, as well as to increase the pumping rate at equal power.

One of the other characteristics of the Multiflo® Trio settler is the presence of an extended settling area arranged between the supply zone and the sludge collection area. Prolonged sedimentation leading to a significant reduction in the amount of sludge admitted between the lamella plates makes it possible to increase the upward velocity (and therefore the flow velocity between the lamella plates) without risking the resuspension of the floc layer deposited on the plates (Figure 5.21). In addition, the large opening between the extended settling area and the sludge recovery zones leads to a reduction in the velocity under the plates and prevents any risk of resuspension of the deposited sludge.

The settling area is equipped with hexagonal modules made up of inclined lamella plates (60°). Between the lamella plates and the level of the sewage sludge bed, there is a 1–2 m zone, which provides a transient regime to promote the separation between water and sludge.

The Multiflo® Trio settler is characterized by the formation of a sludge bed. The flow in the sludge bed is divided evenly in the settler due to coalescence, which continues during this stage and promotes collisions. The sludge flow arriving at the sludge bed does not behave like a piston flow because the low-velocity conditions of the mixture (scraper) reduce the intensity of turbulent fluctuations and create homogeneity, which tends to reduce the bed porosity due to sludge agglomeration.

This homogeneity induces a balance between all the forces present (pressure forces, hydraulic forces), which endows this sludge layer with overall cohesion.

Moderate fluctuations in turbulence continue to promote flocculation by bringing the flocs closer to one another at a minimum distance by means of van der Waals forces.

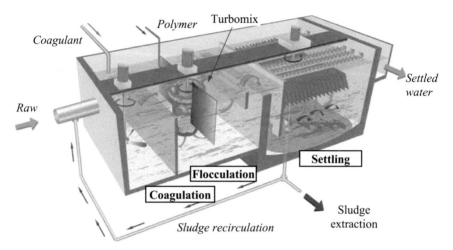

Figure 5.20. *Multiflo® Trio settler. For a color version of this figure, see www.iste.co.uk/gaid/watertreatment1.zip*

One of the interesting consequences is that the sludge bed behaves like a filter in which the new flocculated particles from the flocculation tank are retained on the surface of the flocs already present in the bed. All the more since the flow velocity crossing the sludge bed is calculated in such a way that it disturbs neither the floc agglomeration nor the density of flocs in the bed.

The deepening of the zone under the lamella plates is intended to increase the solids retention time (SRT) to capture a significant part of the suspended solids that failed to be retained in the first zone and to avoid sludge resuspension by the incoming current.

To achieve this, the Multiflo® Trio settler includes equipment making it possible to:

– harrowing of the sludge compartment aiming to remove water pockets and ensure a mixture between the flocs already deposited and the ones arriving from the flocculation tank;

– syncopated extraction of thickened sludge (rate-duration). The duration of extraction is dependent on the mass of sludge produced to avoid the extraction of liquid sludge. Due to the permanent addition of the flocculated colloids and the reagents introduced, the sludge bed's level gradually rises if purges are not carried out regularly to ensure the constant height of this bed. Sludge purges are performed bottom-up without generating dynamic disturbances within the bed.

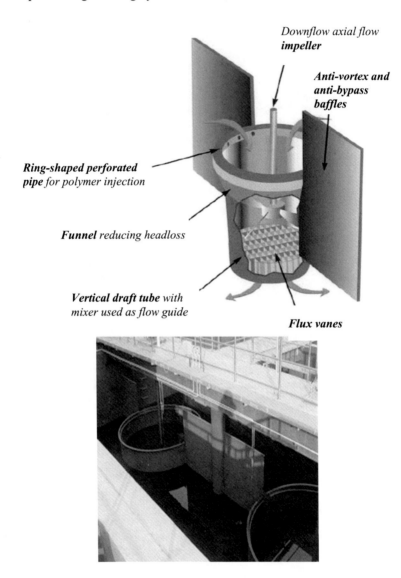

Downflow axial flow **impeller**

Anti-vortex and anti-bypass baffles

Ring-shaped perforated pipe *for polymer injection*

Funnel *reducing headloss*

Vertical draft tube *with mixer used as flow guide*

Flux vanes

Figure 5.21. *Diagram of the Turbomix and installation at the Shanghai-Bailongang plant (China). For a color version of this figure, see www.iste.co.uk/gaid/watertreatment1.zip*

Parameters	Multiflo® Trio
Polymer	Yes
Flocculation	Accelerated
Recirculation circuit	Yes
Maximum mirror velocity	20 m·h^{-1}
Presettling zone	Yes
sludge concentration	Harrow circular scraper
MS sludge concentration	>20 g·L^{-1}
Water loss (<150 NTU)	1%
Flocculation time	10–30' (depending on T°)
Recirculation rate	0-10%
Lamellae gap	30–50 mm
Lamellae length	1–1.5 m
TPA/MS	>15 m^2/m^2
Expected performance (raw water → settled water)	200 NTU → <3 NTU 1,000 NTU → <5 NTU

Table 5.7. *Design parameters for the Multiflo® Trio settler*

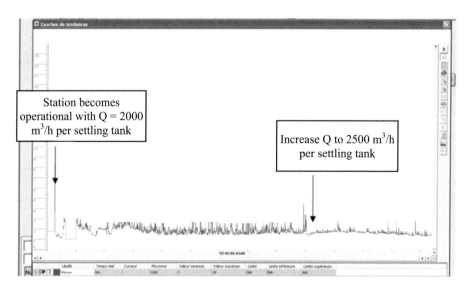

Figure 5.22. *Settled water turbidity at the Crivina plant (Romania) (raw water turbidity variable between 150-4-000 NTU). For a color version of this figure, see www.iste.co.uk/gaid/watertreatment1.zip*

5.4.1.2.1. Performance of the Multiflo® Trio settler

The results on various plants have proven the effectiveness of the Multiflo® Trio settler, as shown in Figure 5.22 obtained on the Crivina plant (Romania). With a mirror velocity of 20 m·h^{-1}, the turbidity of the settled water varies between 1 and 3 NTU, depending on the quality of the raw water. For an inlet turbidity of 4,000 NTU, a 5 NTU turbidity was observed under these operating conditions, that is, a 99.87% efficiency.

5.4.2. Ballasted floc settlers

5.4.2.1. Operating principle

The ballasted floc results from the use of materials introduced during the flocculation phase, which are intended to weigh down and increase the dimensions of the flocs resulting from coagulation. This is evidenced as faster settling. The materials used as ballast can be minerals, either inert or active in nature. Various materials are used: microsand, clays of various kinds, magnetite, activated carbon, etc. Additionally, the sludge (floc) is extracted from the settling stage and is recycled in the flocculation tank.

Mineral materials induce higher settling velocities than recirculated flocs due to their higher density. As a result, this combination increases the settling velocity by a variable factor between 10 and 50, depending on the type of settler (lamella or not). The structures are more compact, and the settled water's turbidity is lower.

Ballasted flocculation is therefore a physicochemical clarification process, functioning at a high velocity and requiring fixation by use of a polymer, flocs or suspended solids on a ballast.

As seen during the flocculation stage, Smoluchowski's expression shows that any increase in particles tends to reduce the contact time:

$$Ln\frac{N}{No} = -\frac{4}{3}\alpha Vp\ Gt$$

where:

– N and No: number of free colloidal particles at time t and to;

– α: effective collision frequency factor;

– Vp: particle volume per volume of suspension;

– G: velocity gradient;

– t: contact time.

A reduction in the Ln N/No ratio leads to high-quality flocculation. This explains that in the event of low turbidity, floc recirculation or particle injection enhances the probability of an encounter. In addition, increasing floc diameter promotes flocculation and increases the chances of encounters between flocs. The specificity of microsand is that it provides a large number of particles (contact surface), improving the flocculation rate, and acting as a ballast, which accelerates floc settling.

Thus, the suspended solids present in drinking water, which were destabilized by the addition of a coagulant, become bound to the microsand particles or to the recycled flocs by means of polymer bridges. These agglomerates of large particles transform into high-density flocs with light mixing energy, and enough retention time encourages the agglomerates to trap random flocs and deliver low turbidity water.

In the presence of microsand—which acts as a ballast—the resulting sludge is collected at the bottom of the settler and is pumped toward hydrocyclones, where the sludge is separated from the microsand by centrifugation. This is then reintroduced into the flocculation tank via the hydrocyclone's underflow.

Figure 5.23. *Ballasted flocculation with microsand promotes rapid settling. For a color version of this figure, see www.iste.co.uk/gaid/watertreatment1.zip*

When the ballast is represented by the concentrated flocs extracted from the settler, the resulting sludge is collected at the settler's bottom and is pumped directly into the flocculation tank.

The ballasted flocculation process has significant advantages, including:

– low floor area;

– fast start-up times to reach peak efficiency in less than 20 min;

– ballasted flocculation can treat a wide range of fluxes with impressive efficiency when removing turbidity and organic substances;

– suitably clarified water in terms of turbidity and suspended solids.

Ballasted flocculation is currently used worldwide for the implementation of coagulation–flocculation–settling.

5.4.2.2. Ballasted floc lamella settlers using recirculated flocs

Flocculation occurs in a floc concentration area maintained at a constant level by means of floc recirculation when the amount of suspended solids in the inlet water is low (< 10 NTU). Recirculation rates vary between 5 and 15%. For this reason, the stirring time and the solid volume concentration in the flocculation zone are greater than those in a conventional flocculator.

The kinetics of the flocculation is theoretically increased, despite a low velocity gradient. Flocs are larger, and the Hazen velocity in the settling area is higher.

Applied to drinking water, these settlers are characterized by an upward velocity between 15 and 18 m·h^{-1}. Note that 1 m and 1.4 m lamella plates are installed.

5.4.2.3. Ballasted floc settlers with inert material (microsand)

Ballasted floc settling with an inert material such as microsand leads to flocculation times of less than 8 min. The mass of microsand crosses the flocculation zone carrying the newly created aggregates due to bridges with the polymer, the metal hydroxides and the suspended matter/colloidal matter (Figure 5.24).

The ballast provided by the microsand injections acts as a primer and initiator for flocculation. The coagulated matter agglutinates around each grain of sand. This material presents a highly developed surface due to its size (60–80 μm) and offers immediate reactivity, even for cold temperatures. The microsand content to feed the flocculation tank should be between 1 and 3 g·L^{-1}. However, it can reach 5–10 g·L^{-1} when the raw water turbidity exceeds 2,000 NTU.

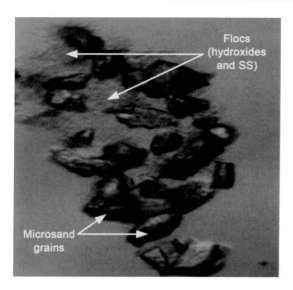

Figure 5.24. *Microsand-floc aggregate (hydroxides-suspended matter/colloidal matter). For a color version of this figure, see www.iste.co.uk/gaid/watertreatment1.zip*

There are two types of Veolia technologies relating to ballasted floc settlers with an inert material:

– Cyclofloc®: ballasted floc settler without lamellae;

– Actiflo®: ballasted floc settler with lamellae.

In both cases, the ballasted floc is based on the injection of a material essentially consisting of microsand.

5.4.2.3.1. Cyclofloc® (ballasted floc settler without lamellae)

Cyclofloc® comes in the form of a tapered concrete tank with a gently sloping floor. Microsand concentrations vary between 1 and 3 $g \cdot L^{-1}$. Coagulated raw water is distributed within a central zone delimited by a cylindrical-conical tank covering almost the entire surface. The flocculant (polymer) is added at the coagulation stage outlet and can also be injected at the level of the conical skirt, located at the center of the cylindrical-conical zone.

The flocculation reaction actually starts in this area. The microsand is also injected at this level so that it reacts both with the flocs created and with the polymer. It is distributed throughout the volume, forming a fluidized bed with flocs (hydroxides-suspended matter-colloidal matter)/polymer. It follows a downward movement while it continues to agglomerate and then settles at the bottom of the structure. The liquid/solid separation then sets in, and the water is evacuated with an upward movement along the perforated tubes on the surface, rejoining a ringed trough. Cyclofloc®'s inclined external slope encourages the still water-suspended flocs to settle down.

No sewage sludge bed is constituted because the microsand/sludge mixture is continually pumped and recirculated before it settles on the floor. A scraper fitted with paddles brings the mixture back to the center of a pit from which it is extracted by means of a recycling pump. This sends the sand-sludge mixture back to a hydrocyclone, which restores the microsand underflow within the conical skirt. The sludge extracted in overflow mode is sent to sludge treatment.

The hydrocyclone is an elongated cylindrical-conical chamber, with an end pointing downward. The sand-sludge mixture arrives to the cylindrical section under pressure and tangentially. Heavy particles such as sand are pushed toward the periphery due to the action of centrifugal forces.

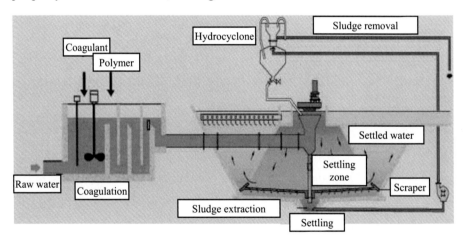

Figure 5.25. Cyclofloc® diagram. For a color version of this figure, see www.iste.co.uk/gaid/watertreatment1.zip

The upper orifice is placed in such a way that it only lets out the liquid portion drawn from the center, microsand-free. At the lower end of the hydrocyclone, the microsand is recollected for reinjection.

The mirror velocity rate is between 8 and 10 m·h⁻¹, resulting in an effluent with an average turbidity of less than 2.5 NTU for inlet turbidities reaching 500 NTU.

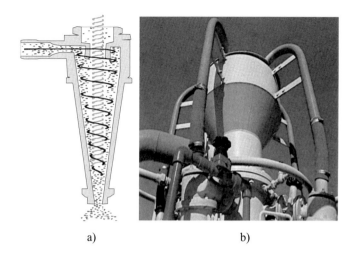

a)   b)

Figure 5.26. *(a) Hydrocyclone and (b) hydrocyclone battery. For a color version of this figure, see www.iste.co.uk/gaid/watertreatment1.zip*

Figure 5.27. *Drinking water plants using Cyclofloc® (Gour du Puy (a), Villeneuve-sur-Lot (b), Orleans (c) and Douai (d)). For a color version of this figure, see www.iste.co.uk/gaid/watertreatment1.zip*

5.4.2.3.2. Actiflo® (ballasted floc lamella settler)

Operating principle

The operating principle of Actiflo® is similar to that of Cyclofloc®, with injected microsand acting as flocculation ballast and microsand recirculation in the flocculation tank.

Actiflo® has a coagulation zone, a microsand injection zone, a maturation zone and finally, a settling area (Figure 5.28).

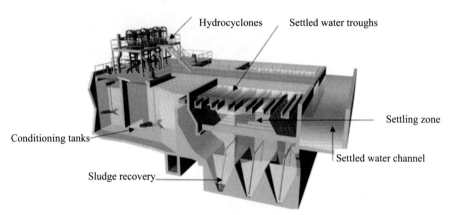

Figure 5.28. *Actiflo® operating principle. For a color version of this figure, see www.iste.co.uk/gaid/watertreatment1.zip*

Coagulation–injection–maturation

Raw water enters a coagulation tank into which the coagulant is injected, which destabilizes colloidal matter. The coagulant can also be injected directly upstream (in-line injection). The contact time is 2 min but can rise to 3–4 min at very low temperatures. The velocity gradient G is between 400 and 700 s^{-1}.

The coagulated water then passes into an injection tank into which the recirculated microsand and the polymer are injected. Rapid mixing ensures a dispersion of microsand, whereas the polymer enables microflocs and other suspended matter to become attached to the microsand. The grain size of the microsand is between 60 and 80 μm. The reactivity of its large specific surface triggers agglomeration reactions with the solids present (microflocs and suspended matter) and the polymer (Figures 5.24 and 5.29). The contact time is usually 2 min.

The mixture then passes into a third tank, called the maturation tank, where flocs mature. The retention time was 6 min. The velocity gradient G is between 120 and

165 s^{-1}. Finally, liquid/solid separation takes place in the countercurrent lamella settling tank.

The settled water is evacuated by a collector toward the filtration stage. The decanted flocs are collected either by a scraper toward a central sump or at the bottom of a tremie. The sediment formed from sludge and microsand is continuously pumped to a hydrocyclone. The hydrocyclone creates a centrifugal vortex that separates the microsand and sludge. In this way, almost all of the microsand is recovered. Specific equipment makes it possible to control the reagent dosages (coagulant and flocculant), as well as the concentration of microsand in the recirculation flow, the recycling rate and the velocity gradient in the coagulation and flocculation tanks.

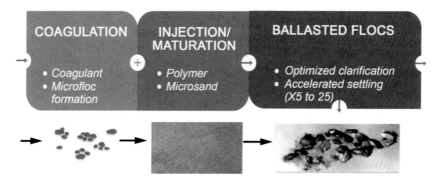

Figure 5.29. *Ballasted flocculation principle in Actiflo$^®$. For a color version of this figure, see www.iste.co.uk/gaid/watertreatment1.zip*

The Stokes equation shows that for a 300 μm diameter of agglomerated flocs, the flocs' falling velocity rises to 81 m·h^{-1}. This means that Actiflo$^®$ can operate in drinking water at velocitys of at least 60 m·h^{-1}

$$Vc = \frac{g\, d^2(\rho s - \rho w)}{18\, \mu}$$

where:

– g: gravitational constant (9.81 m^2·s^{-1});

– μ: dynamic viscosity (1.13 × 10^{-3} kg·m^{-1}·s^{-1});

– d: floc diameter (m);

– ρs: density of agglomerated flocs (kg·m^{-3});

– ρw: water density (kg·m^{-3}).

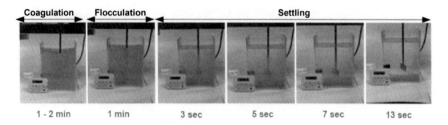

Figure 5.30. *Actiflo® jar test. For a color version of this figure, see www.iste.co.uk/gaid/watertreatment1.zip*

The most important advantage of the settler is its retention time, which is less than 15 min for a nominal design flow. This time is significantly shorter than that in a conventional process. The use of ballasted flocculation therefore requires much less space. The space required takes up approximately 5% of the space used by a conventional settling system. This technology makes it possible to produce water with qualities equal to or superior to the water produced using conventional processes and using a much smaller space.

The ballasted flocculation mechanisms with microsand can be explained using the ratio:

$$\frac{dN}{dt} = -KaGN + KbG^2$$

where:

– N: particle concentration;

– Ka: aggregation constant;

– Kb: deterioration constant (s);

– G: velocity gradient (s^{-1});

– t: contact time (s).

The first term of the equation is a measure of the extent of aggregation, whereas t he second term measures the extent of aggregate deterioration due to turbulence, excessively violent collisions, and temperature. The results show that in the case of Actiflo, the optimal velocity gradient is approximately three to ten times higher (150–500 s^{-1}) than that involved in a conventional process (50 s^{-1}). The deterioration constant Kb is 25 times smaller than that for a conventional treatment. This ratio shows that the aggregation forces are significantly higher than the flocs' deterioration and shear forces.

The results obtained show that the flocs formed are more solid and cohesive. They can resist an increase in the velocity gradient, which is necessary due to the microsand mass in the reaction tank.

The flow is laminar and unidirectional. The floc is subject to only two forces: the feed rate velocity (rising velocity) and gravitation. There is a strong interaction between the particles, and there is no influence due to the shape of flocs (or other particles present in the water) because gravitation is conditioned and managed by the mass of microsand present in the mixture.

The decanted mixture (sludge and microsand) is pumped continuously to a hydrocyclone that separates the microsand and the sewage sludge. The microsand is recycled in the injection tank by the hydrocyclone's underflow, while the sludge is evacuated in overflow mode.

The microsand concentration ($kg \cdot m^{-3}$) is calculated according to the equation

$$m = \frac{1.5 \times Qr}{Qw}$$

Based on the density of the microsand in the mixture equal to 1.5 $kg \cdot m^{-3}$, the recirculation flow rate Qr ($m^3 \cdot h^{-1}$) oscillates between 3 and 12%. Qw is the raw water inlet flow rate in Actiflo ($m^3 \cdot h^{-1}$).

Overflow rate = 80%, $FeCl_3$ = 25 $mg \cdot L^{-1}$

With microsand separation efficiency normally greater than 99.9% and a concentration in the recirculation line of approximately 100–150 $g \cdot L^{-1}$ (for a sand treatment dose of 3 $kg \cdot m^{-3}$ at nominal flow with a recirculation rate of 3%), it results in a 3 $g \cdot m^{-3}$ maximum loss of sand-sludge in the water to be treated.

The overall sand loss on Actiflo® also includes microsand leaks into the settled water, which depend on the quality of flocculation and the settling rate. It is sometimes compensated by a non-negligible contribution of microsand from raw water, particularly in the case of rivers, or by a make-up of new microsand.

The choice of hydrocyclone essentially depends on the flow to be cycloned and the grain size of the sand used.

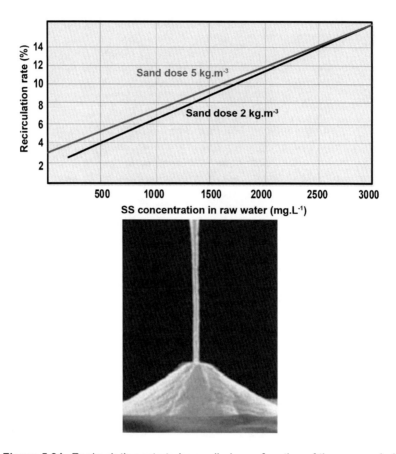

Figure 5.31. *Recirculation rate to be applied as a function of the suspended solids content in raw water. For a color version of this figure, see www.iste.co.uk/gaid/watertreatment1.zip*

The choice of coagulant depends on the characteristics of the water to be treated, the treatment's goal (removal of turbidity, thorough removal of organic matter, etc.) and pH constraints. The doses to be applied should be defined and optimized by means of jar tests.

The amount of flocs only generated by reagent dosages depends on a coefficient K related to the hydroxides bound to the coagulant (see section 2.3).

The sand loss with sludge (in hydrocyclone overflow) varies depending on the effective size and dose of sand, the recirculation rate, the operating conditions of hydrocyclones and the sludge concentration to be evacuated.

Ballasted floc settling rate

For a floc-sand aggregate of 200 μm, the settling rate (vp) deduced from the Stokes equation is given as:

$$vp = \frac{g(\rho s - \rho w)dp^2}{18\eta}$$

$$vp = \frac{9 \cdot 81 \text{ m} \cdot \text{s}^{-2}(2.6 - 1)(200 \times 10^{-6})^2}{18 \times 1.003 \times 10^{-6} \text{m}^2 \cdot \text{s}^{-1})} = 0.0348 \text{ m} \cdot \text{s}^{-1}$$

$$= 125 \text{ m} \cdot \text{h}^{-1}$$

where:

− ρs: sand density: 2.6 g·cm^{-3}, ρw density of water: 1 g·cm^{-3}, ballasted flocs diameter: 200 μm: 200×10^{-6} m;

− η: kinematic viscosity: 1.003×10^{-6} m²·s^{-1}.

Reynolds number is given as:

$$Re = \frac{\Phi vpdp}{\eta} = \frac{(2.5)\,(0.0348)\,(200 \times 10^{-6})}{1.003 \times 10^{-6} \text{m}^2 \cdot \text{s}^{-1}} = 17.3$$

where Φ is the shape factor: 2.5.

Since Re > 1, Newton's law must be applied to assess the settling rate in the transition zone.

The drag coefficient C_D is given by:

$$CD = \frac{24}{Re} + \frac{3}{\sqrt{Re}} + 0.34$$

That is:

$$CD = \frac{24}{17.3} + \frac{3}{\sqrt{17.3}} + 0.34 = 2.445$$

We then have

$$vp = \sqrt{\frac{4g(\rho s - \rho w)d}{3\,CD\Phi}} = \sqrt{\frac{4(9.81)(2.6-1)(200 \times 10^{-6})}{3\,(2.445)(2.5)}} = 0.026 \text{ m·s}^{-1} = 93.6 \text{ m·h}^{-1}$$

The velocity values obtained in the two cases are not identical. Supplementary iteration is necessary. Assuming a new settling rate of 0.022 m.s^{-1}, the corrected Reynolds number is given as

$$Re = \frac{\Phi v p d p}{\eta} = \frac{(2.5)\,(0.022)\,(200 \times 10^{-6})}{1.003 \times 10^{-6}\,\text{m}^2\cdot\text{s}^{-1}} = 10.97$$

$$CD = \frac{24}{10.97} + \frac{3}{\sqrt{10.97}} + 0.34 = 3.43$$

$$vp = \sqrt{\frac{4g(\rho s - \rho w)d}{3\,CD\Phi}} = \sqrt{\frac{4(9.81)(2.6-1)(200 \times 10^{-6})}{3\,(3.43)(2.5)}} = 0.022 \text{ m}\cdot\text{s}^{-1} = 79.5 \text{ m}\cdot\text{h}^{-1}$$

Actiflo® basic design and performance

Basic design

Table 5.8 presents the basic design for Actiflo®.

Mirror velocity Vm	m·h^{-1}	40	60	90*
Coagulation retention time	min	2	2	1.33
Injection retention time	min	2	2	1.33
Maturation retention time	min	6	6	4

* Under certain raw water qualities and operating conditions.

Table 5.8. Actiflo® basic design

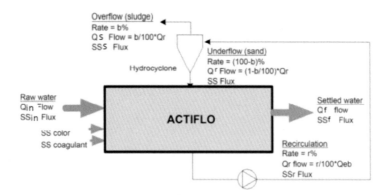

Figure 5.32. Thorough assessment of suspended solids and sewage sludge on Actiflo®. For a color version of this figure, see www.iste.co.uk/gaid/watertreatment1.zip

The following evaluation makes it possible to determine the concentration of suspended solids in the sludge extracted and the recirculated sand, depending on the quality of the inlet water and the treatment applied.

Example of design of Actiflo® settler: inlet flow: 500 $m^3 \cdot h^{-1}$, and T °C: 15°C.

Parameters	Coagulation	Injection	Maturation
Number of tanks	1	1	1
Reaction time (min)	2	2	6
Total volume (m^3)	18.4	18.4	54.7
Water level (m)	4.6	4.6	4.6
Total projected area (m²)	4	4	11.9
Unit length (m)	2	2	3.45
$G = ((P/\mu.V)1/2)$ (s^{-1})	366	366	174
Impeller power (W)	2,465	2,465	1,667
μ dynamic viscosity ($kg \cdot m^{-1} \cdot s^{-1}$)	0.001	0.001	0.001
Impeller velocity (min^{-1})	110	110	25
Np power number	0.4	1	1
ρ water ($kg \cdot m^{-3}$)	1,000	1,000	1,000
D paddle diameter (m)	1.00	1.00	1.80
Gt	44,930	44,930	62,845

Settling	
Parameters	Values
Settler unit	1
Mirror surface (m²)	12.5
Unit width (m)	2.55
Unit length (m)	4.90
Non-effective length	0.5
Total length (m)	5.40
Lamella length (m)	1.00
Retention time (min)	11.1
Settling volume (m^3)	92.5
Gap (mm)	36
TPA/mirror s	15.38
Total projected area	192.3
Hazen velocity $(m \cdot h)^{-1}$	4.8
Mirror velocity $(m \cdot h^{-1})$	40

Table 5.9. *Design parameters of the Actiflo® settler*

Performance: turbidity removal

Table 5.10 presents a comparison between the turbidity removal results on Actiflo® and on a conventional lamella settling tank. It appears that while the results obtained on conventional settlers are acceptable, those obtained on Actiflo® are clearly better, often requiring lower reagent dosages, regardless of the inlet turbidity.

Country	Raw water turbidity (NTU)	Total organic carbon (mg·L^{-1})	Types of coagulant	Coagulants		Actiflo	
				Coagulant dose (g·m^{-3})	Settled water turbidity (NTU)	Coagulant doses (g·m^{-3})	Settled water turbidity (NTU)
France	5–20	3–5	Al$_2$(SO$_4$)$_3$	25–45	<1.5	20–40	0.5–0.8
France	<5	<2	Al$_2$(SO$_4$)$_3$	6–7	1–1.5	5–6	0.4–0.5
France	25–40	4–6	FeCl$_3$	55–60	1.5–2	55–60	0.7–1.1
Great Britain	<5	3	Al$_2$(SO$_4$)$_3$	20	0.8–1.2	15–20	0.3–0.6
China	30–60	<3	PACl	–	0.6–0.7	–	0.6–0.7
Canada	4–18.7		Al$_2$(SO$_4$)$_3$	10–12	0.5–0.7	6–8	0.2–0.5
Malaysia	500–3,000	<3	Al$_2$(SO$_4$)$_3$	–	–	45	<2
US	50–90	–	Fe$_2$(SO$_4$)$_3$	14–15	3–5	10–15	0.6–0.7
US	3–10	–	Al$_2$(SO$_4$)$_3$	6–7	1–2	5–6	0.4–0.5

Table 5.10. *Actiflo performances for turbidity removal*

The turbidity parameter monitoring curves show the performance of ballasted sedimentation using microsand for raw water turbidity between 20 and 45 NTU and for higher turbidity between 20 and 1,600 NTU. The turbidity of the Selangor River (Malaysia) rose to more than 4,500 NTU, and the turbidity of the Actiflo settled water remained below 5 NTU.

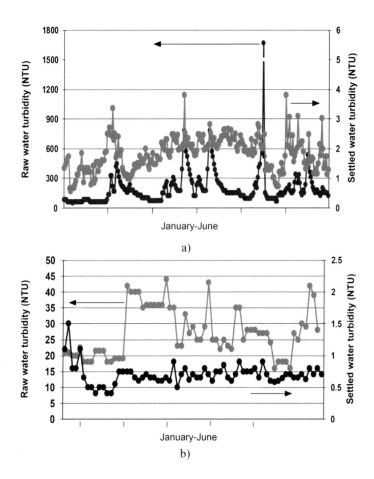

Figure 5.33. *Turbidity removal in two surface waters (Malaysia (a) and France (b)). For a color version of this figure, see www.iste.co.uk/gaid/watertreatment1.zip*

Not only does color removal (Figure 5.34) depend on the dosage and choice of reagents but also on pH. The results obtained with Actiflo® on various plants whose raw water has low turbidity (<10 NTU) and a high true color concentration (up to 400 mg·L^{-1} Pt/Co), show color values between 5 and 15 mg·L^{-1} Pt/Co.

Actiflo is highly effective for removing algae. They are trapped at the same time as colloidal particles and other particles in suspension. Figures 5.35 and 5.36 properly illustrate this effectiveness: jar tests carried out on algae concentrations of several thousand cells·mL^{-1} result in an efficiency of more than 99%.

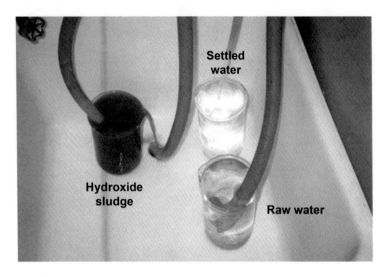

Figure 5.34. *Color removal. For a color version of this figure, see*
www.iste.co.uk/gaid/watertreatment1.zip

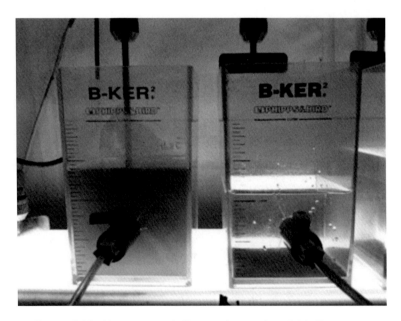

Figure 5.35. *Algae removal. For a color version of this figure, see*
www.iste.co.uk/gaid/watertreatment1.zip

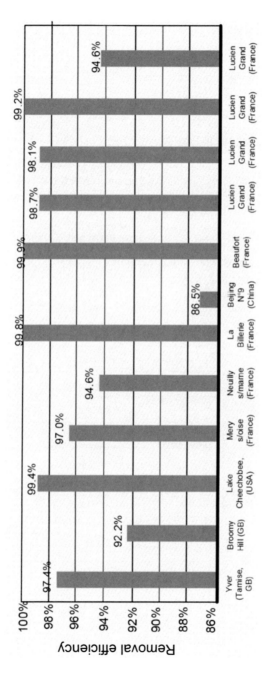

Figure 5.36. *Algae removal on Actiflo®. For a color version of this figure, see www.iste.co.uk/gaid/watertreatment1.zip*

The results obtained in drinking water plants show that the lowest efficiency is 86.5% and that efficiency values approaching or exceeding 98% are frequent (Figure 5.36).

Several versions of Actiflo® have been applied to drinking water, such as the following:

– Actiflo® HCS, which reduces the sludge volume;

– Actiflo® Softening, including caustic soda or lime, which reduces water hardness (see Volume 3, Chapter 15);

– Actiflo® Carb, including activated carbon, favors the removal of organic matter, pesticides and micropollutants (drugs, endocrine disruptors, etc.) (see Volume 3, Chapter 10);

– Actiflo® Rad, which removes radioactivity particles;

– Actiflo® Turbo, which reduces contact times thanks to a turbo-mix installed in the flocculation tank.

Table 5.11 shows the overall performance of Actiflo® in relation to many parameters applied to drinking water.

Conventional surface water			
Parameters	**Removal, %**	**Parameters**	**Elimination, %**
Turbidity	90->99%	**Fe-Mn**	>90%
SS	90->99%	**Softening**	>40 °F*
Color	80–95%	**Parasites** (*Cryptosporidium–Giardia*)	>2.9 log
DOC	40–70%	**Bacteria**	>2 log
Cts particles (**2–15 μm**)	>2 log	**Viruses**	>2 log
Algae	85->99%	**Helminth eggs**	>1.5 log

*See Volume 3, Chapter 15.

Table 5.11. *Actiflo® performance on certain parameters*

5.4.2.3.3. Actiflo Turbo

The Actiflo® Turbo features the installation of a Turbomix within the flocculation tank. This equipment has a large number of advantages such as:

– the removal of dead zones, thus increasing the useful volume;

– removal of the hydraulic by-pass;

– mechanical protection of paddles and therefore of the impeller by deflecting the incoming flow;

– an increase in the pumping rate by converting a radial flow into an axial flow;

– ensures proper mixing in the tanks at shallow depth, preventing liquid rotation (vortex);

– the use of single-stage impellers in deep tanks;

– improves the distribution of retention times by shortening them and therefore also reducing tank volumes;

– produces more homogeneous flocs;

– the removal of the injection tank compared to Actiflo®;

– impeller downsizing to fit a smaller flocculation tank.

Actiflo® Turbo includes a 2-min coagulation stage, followed by a flocculation stage, which includes the microsand and polymer injection, as well as the maturation stage. This stage implies a retention time of 4 min. The value of Gt is equal to 36,000.

Liquid–solid separation takes place in the lamella settling, following the flocculation stage (Figure 5.37). The operation of the Actiflo® Turbo is similar to that of the conventional Actiflo®. Actiflo® Turbo is applicable for clearly defined raw water qualities and operating conditions. Veolia has built over a hundred Actiflo Turbos worldwide.

Veolia has more than 1,100 Actiflo® references installed for the treatment of drinking water, municipal wastewater (including rainwater), process water and industrial wastewater.

Veolia has developed a complete range of prefabricated Actiflo® units and standardized racks for the treatment of drinking water, process water and municipal and industrial wastewater. Depending on the configuration and application, the processing capacity per unit can reach up to 2,500 $m^3 \cdot h^{-1}$.

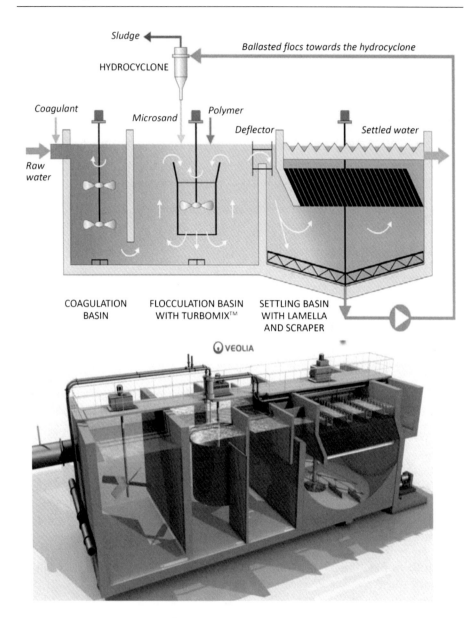

Figure 5.37. *Actiflo*® *Turbo operating principle. For a color version of this figure, see www.iste.co.uk/gaid/watertreatment1.zip*

Figure 5.38. *Drinking water plants working with Actiflo®: (a) Lucien Grand (France), (b) Oslo (Norway), (c) Quebec (Canada), (d) Beijing No. 9 (China). For a color version of this figure, see www.iste.co.uk/gaid/watertreatment1.zip*

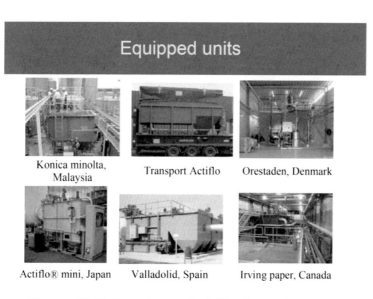

Figure 5.39. *Units equipped with Actiflo. For a color version of this figure, see www.iste.co.uk/gaid/watertreatment1.zip*

5.5. References

Al-Dulaimi, S.M.S. and Racoviteanu, G. (2018). Performance of the tube settler clarification at different inclinaison angles and variable flow rate. *Mathematical Modelling in Civil Engineering*, 14(2), 13–25.

Argaman, Y. and Kaufman, W.F. (1970). Turbulence and flocculation. *J. Sanit. Eng. Div., Proc. Am. Soc. Civ. Eng.*, 96, 226.

Camp, T.R (1953). Flocculation and flocculation basins. *Proc. ASCE*, 1, 79–89.

Camp, T.R. and Stein, P.C. (1943). Velocity gradients and internal work in fluid motion. *J. Boston Soc. Civ. Eng.*, 10(30), 219–237.

Camp, T.R., Rost, D.K., Bhosta, B.V. (1940). Effects of temperature on the rate of floc formation, *J. Amer. Water Works Assoc.*, 32, 913–927.

Cantwell, R.E. and Hofmann, R. (2011). Ultraviolet absorption properties of suspended particulate matter in untreated surface waters. *Water Research*, 45(3), 1322–1328.

Crittenden, J.C., Rhodes Trussell, R., Hand, D.W., Howe, K.J., Tchobaoglous, G. (2012). *Water Treatment Principles and Design*, 3rd ed. John Wiley & Sons, New York.

Culp, G., Hansen, S., Richardson, G. (1968). High-rate sedimentation in water treatment works. *J. Amer. Water Works Assoc.*, 60(6), 681–698.

Davis, M.L. (2010). *Water and Wastewater Engineering, Design Principles and Practice*. McGraw Hill, New York.

De Dianous, F. and Dernaucourt, J. (1991). Advantages of weighted flocculation in water treatment. *Water Supply*, 9, 43–46.

Desjardins, C., Koudjonou, B., Desjardins, R. (2002). Laboratory study of ballasted flocculation. *Water Research*, 36(3), 744–754.

Demir, A. (1995). Determination of settling efficiency and optimum plate angle for plated settling tanks. *Water Research*, 29, 611–616.

Demir, S. (2014). Practical model for estimating pressure drop in cyclone separators: An experimental study. *Powder Technology*, 268, 329–338.

Fadel, A.A. and Baumann, E.R. (1990). Tube settler modeling. *Am. Soc. Cir. Engrs*, 116, 107–124.

Fair, G.M., Geyer, J.C., Okun, D.A. (1968). *Water and Wastewater Engineering*, volume 2. John Wiley & Sons, New York.

Farrell, C., Hassard, F., Jefferson, B., Leziart, T., Nocker, A., Jarvis, P. (2018). Turbidity composition and the relationship with microbial attachment and UV inactivation efficacy. *Science of the Total Environment*, 624, 638–647.

Fitzpatrick, C.S.B., Fradin, E., Gregory, J. (2004). Temperature effects on flocculation, using different coagulants. *Water Science and Technology*, 50(12), 171–175.

Forsell, B. and Hedström, B. (1975). Lamella sedimentation: A compact separation technique. *Water Pollution Control Federation*, 47(4), 834–842.

Ghanem, A., Young, J., Edwards, F. (2007). Mechanisms of ballasted floc formation. *J. Environ. Eng.*, 133(3), 271–277.

Goula, A.M., Kostoglou, M., Karapantsios, T.D., Zouboulis, A.I. (2008). A CFD methodology for the design of sedimentation tanks in potable water treatment case study: The influence of a feed flow control baffle. *Chem. Eng. J.*, 140, 110–121.

Gregory, J. and Barany, S. (2011). Adsorption and flocculation by polymers and polymer mixtures. *Adv. Colloid Interface Sci.*, 169(1), 1–12.

Guibelin, E., Delsalle, F., Binot, P. (1994). The Actiflo process: A highly compact and efficient process to prevent water pollution by stormwater flows. *Water Sci. Technol.*, 30(1), 87–96.

Haarbo, A., Dahl, C., Rineau, S. (1998). Successful applications: New applications of the Actiflo process. *Phy. Chem. Process, Water Quality International*, 29–30.

Hazen, A., Pearsons, G.W., Weston, R.S., Fuller, G.W. (1904). Discussions on sedimentation. *J. Assoc. Eng. Soc.*, 53, 45–88.

Hilal, N. and Johnson, D. (2019). Interaction between ballasting agent and flocs in ballasted flocculation for the removal of suspended solids in water. *Journal of Water Process Engineering*, 33, 101028.

Imam, E., McCorquodale, J.A., Bewtra, J.K. (1983). Numerical modeling of sedimentation tanks. *J. Hydraul. Eng. ASCE*, 109, 1740–1754.

James, C.Y. and Edwards, G.F (2003). Factors affecting ballasted flocculation reactions. *Water Environment Research*, 75(3), 263–272.

Kang, L.S. and Cleasby, J.L. (1995). Temperature effects on flocculation kinetics using Fe(III) coagulant. *J. Environ. Eng.*, 121(12), 893–901.

Kynch, C.J. (1952). A theory of sedimentation. *Trans. Faraday Soc.*, 48, 166–171.

Lapointe, M. and Barbeau, B. (2015). Evaluation of activated starch as an alternative to polyacrylamide polymers for drinking water flocculation. *J. Water Supply Res. Technol. – Aqua*, 64(3), 333–343.

Lapointe, M. and Barbeau, B. (2018). Selection of media for the design of ballasted flocculation processes. *Water Res.*, 145, 25–32.

Larsen, P. (1977). *On the Hydraulics of Rectangular Settling Basins*. Lund Institute of Technology, Lund.

Lengo, K.M. (1994). Compared effects of various coagulants on organic matter removal from drinking waters – Influence of prehydrolysation. PhD Thesis, École Polytechnique de Montréal, Montreal.

Maulding, J.S. and Harris, R.H. (1968). Effect of ionic environment and temperature on the coagulation of colour-causing organic compounds with ferric sulphate. *J. Amer. Water Works Assoc.*, 60(4), 60–76.

McCorquodale, J.A. and Zhou, S. (1993). Effects of hydraulic and solids loading on clarifier performance. *J. Hydraul. Res.*, 31, 461–477.

Moharram, F. and Shaban, H. (2018). The performance of sludge blanket clarifier against conventional settler under high water turbidity conditions. *Water Practice and Technology*, 13(3), 642–653.

Pujol, E., Vullierme, M., Druoton, J.C., Sibony, J. (1993). Performances of the Actiflo clarifier at the Neuilly-sur-Marne treatment plant. *TSM*, 11, 457–462.

Sibony, J. (1981). Clarification within microsand seeding. A state of the art. *Water Research*, 15, 1281.

Smoluchowski, M.V. (1917). Versuch einer mathematischen Theorie der Koagulationskinetik kolloider Losungen. *Z. Phys. Chem.*, 92(9), 129–168.

Takata, K. and Kurose, R. (2017). Influence of density flow on treated water turbidity in a sedimentation basin with inclined plate settler. *Water Science & Technology: Water Supply*, 17(4), 1140–1148.

Tarpagkou, R. and Pantokratoras, A. (2014). The influence of lamellar settler in sedimentation tanks for potable water treatment – A computational fluid dynamic study. *Powder Technology*, 268, 139–149.

Thomas, D., Judd, S., Fawcett, N. (1999). Flocculation modelling: A review. *Water Res.*, 33(7), 1579–1592.

Tikhe M.L. (1974). Some theoretical aspects of tube settlers. *Ind. J. Environ. Health*, 16, 26–33.

Van Benschoten, J.E. and Edzwald, J.K. (1990a). Chemical aspects of coagulation using aluminum salts I. Hydrolytic reactions of alum and polyaluminum chloride. *Water Res.*, 24(12), 1519–1526.

Van Benschoten, J.E. and Edzwald, J.K. (1990b). Chemical aspects of coagulation using aluminum salts II. Coagulation of fulvic acids using alum and polyaluminum chloride. *Water Res.*, 24(12), 1527–1535.

Van Benschoten, J.E., Edzwald, J.K., Rahman, M.A. (1992). Effects of temperature and pH on residual aluminium for alum and polyaluminium coagulants. *Water Supply*, 10(4), 49–54.

Vesga-Rodríguez, C.P., Donado-Garzón, L.D., Weber-Shirk, M. (2019). Evaluation of high rate sedimentation lab-scale tank performance in drinking water treatment. *Revista Facultad de Ingeniería, Universidad de Antioquia*, 90, 9–15.

Willis, R.M. (1978). Tubular settlers – A technical review. *J. Amer. Water Works Assoc.*, 70(6), 331–335.

Yao, K.M. (1973). Design of high-rate settlers. *Proc. Am. Soc. Civil Eng.*, 99(EE5), 621–632.

Young, J.C. and Edwards, F.G. (2000). Fundamentals of ballasted flocculation reactions. *Proc. Water Environ. Fed.*, 14, 56–80.

Young, J.C. and Edwards, F.G (2003). Factors affecting ballasted flocculation reactions. *Water Environment Research*, 75(3), 263–272.

Yukselen, M.A. and Gregory, J. (2004). The reversibility of floc breakage. *Int. J. Miner. Process.*, 73(2–4), 251–259.

Zioło, J. (1996). Influence of the system geometry on the sedimentation effectiveness of lamella settlers. *Chemical Engineering Science*, 51(1), 149–153.

6

Flotation

Flotation is a single operation used for liquid–solid separation through the injection of air bubbles into the system, generating particle–bubble agglomerates with sufficient buoyancy against a downward flow. If buoyancy is not enough, the agglomerate is carried toward the treated water outlet.

The attachment of gas bubbles to the surface of the dispersed phase is only possible if the gas–solid affinity is greater than the liquid–solid affinity. When a solid surface is intimately associated with the surrounding liquid, as is the case for hydrophilic particles for example, there is no precise interface. The gas cannot displace the liquid on the solid's surface because it is unable to access it. For this, it is essential to reverse such an effect by means of flotation. In the presence of particles that can become wet, their condition must be modified by covering their surface with a molecule film that has little affinity with the liquid. This results in a reduction in the "solid–liquid" adhesion forces and an increase in the "solid–gas" forces. A solid's buoyancy is considered all the better the less it allows itself to be made wet by the liquid.

In water, solids are classified into:

– polar substances (molecules with polarized bonds). Polar substances are hydrophilic. The water molecule, itself polar, is sensitive to the electric field created by the ions from ionic compounds;

– non-polar substances (molecules with covalent bonds), which are hydrophobic.

If the particles brought into contact with the bubbles are captured by these, they can rise to the cell's surface, where these agglomerates become concentrated in the form of sludge, which will then be evacuated from the structure in the form of discharge.

Fine air bubbles are injected into a contact zone where solid and flocculated particles mix up with the bubbles to create a particle–bubble aggregate. Air bubbles become attached to the solids by selective fixation on their surface and, in turn, make the particles rise to the surface.

This process exploits the density difference between the two phases present. Several cases can be differentiated:

– if the density of the solid phase is significantly lower than that of the liquid phase, the separation is said to be natural (example of microalgae);

– if the density of the solid phase is slightly lower than that of the liquid phase, the flotation is said to be assisted, because external means must be implemented to improve the separation;

– if the density of the solid phase, originally greater than that of the liquid phase, is artificially reduced, flotation is said to be induced (example of flocs).

The main advantage of dissolved air flotation (DAF) is that particles, which are less dense than the surrounding water, are more effectively removed by flotation rather than by a conventional sedimentation process.

6.1. The scope of DAF

DAF technology is particularly recommended for the removal of clogging particles such as algae or other solid particles that cannot be easily removed by settling or direct filtration when their concentration is too high.

In general, DAF systems are installed for the treatment of water either with low turbidity but rich in algae or which are highly colored. However, DAF can work with turbidities of up to 50 NTU.

DAF systems can produce treated water <1 NTU and can be combined with gravity filtration (monolayer or bilayer) or membrane filtration methods.

The velocity applied in DAF systems is two to three times higher than that of lamella settlers operating at a high velocity. In some special cases, they compete with ballasted floc settlers with an inert material.

The concentration of extracted sludge is much higher than that of settlers. It is between 2 and 3%.

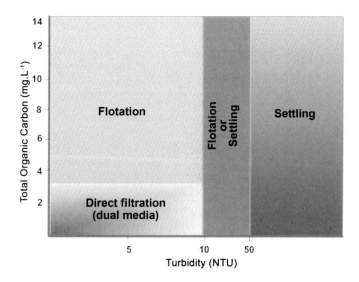

Figure 6.1. *The role of flotation compared to other processes. For a color version of this figure, see www.iste.co.uk/gaid/watertreatment1.zip*

6.2. The main stages of a flotation process

Very small colloidal particles present removal difficulties that are specific to them. They are the seat of incessant shocks due to Brownian motion and do not agglomerate due to the existence of repulsive electrical charges on their surface, which contribute to their stability. The polarity they acquire is due to the adsorption of cations or anions (positive or negative colloids). The agglomeration of these particles among themselves or their flocculation requires breaking up their means of protection, in other words, charge neutralization.

In general, a flotation process comprises a coagulation, a flocculation and a flotation stage (Figure 6.2). This is divided into two adjacent compartments: a contact zone where the injection of air bubbles from supersaturated water promotes the formation of a floc–bubble aggregate and a solid–liquid separation zone.

6.2.1. *Coagulation*

Coagulation is the physical process whereby colloidal particles are destabilized to facilitate their aggregation. This destabilization is obtained by adding a chemical coagulant, such as a metal salt or a polymerized metal coagulant. The choice of coagulant depends on the quality of the water to be treated.

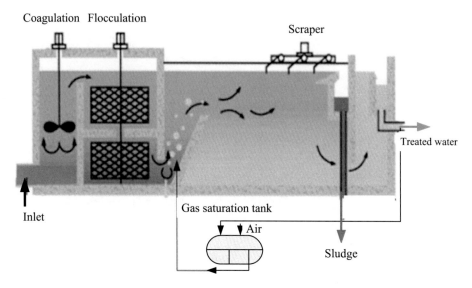

Figure 6.2. *Operating principle of a flotation unit. For a color version of this figure, see www.iste.co.uk/gaid/watertreatment1.zip*

6.2.2. Flocculation

During this stage, destabilized colloid particles agglomerate to form flocs. The flocculation time is between 10 and 20 min, depending on the quality of the water to be treated and the temperature.

6.2.3. Flotation

Flocculated water arrives in the flotation zone, which comprises the contact zone and the separation zone.

6.2.3.1. Contact zone

Flocculated raw water is distributed toward the bottom of the contact zone, following an upward movement. In this zone, there is a floor covered with specific nozzles that are fed with supersaturated water. This water is prepared from good quality floated water or from filtered water. It is placed under pressure (3–6 bars) in a gas saturation tank and applied with a recycling pump. This water flow is often referred to as "recirculation flow". Due to the pressure drop of supersaturated water at the nozzles, small air bubbles are formed. The number of bubbles, as well as their mean size and size distribution, are important parameters for the separation efficiency of the flotation process. The rising velocity of air bubbles is greater than the water velocity, entailing the collision of air bubbles, which in turn become attached to the flocs by fixing up on their surface. The density of floc–bubble aggregates decreases to values lower than the water density, which causes aggregates to float on the water surface. The contact time was 1–2 min. The details of the mechanisms taking place in this zone will be described later. The recirculation rate varies between 5 and 10% but can reach values of 20% when the concentration of particles in water is very high. This is what happens during the periods of algal bloom, for example.

6.2.3.2. *Separation zone*

In the separation zone, the aggregation of flocs and air bubbles leads to the formation of floating sludge on the floating unit's surface. This concentrated sludge is drained using a scraper or by discharging its overflow toward the sludge treatment zone. The water thus produced is collected at the separation zone's floor level, with exclusively adapted perforated tubes.

6.3. The fundamental mechanisms of flotation

The bubble–particle interaction mechanisms control the efficiency of the flotation process. These mechanisms can be divided into three stages.

At first, a bubble and a particle approach one another until they form a liquid film between them. This is the collision stage. It is during this stage that surface forces come into play. Several mechanisms are involved:

– Interception: small particles adapt to the flow, and their trajectory is identical to the streamline. Due to the streamlines' compression as the bubble moves by, these particles are brought to the bubble's surface due to the liquid's movement.

– Inertia: the greater the particle's inertia, the more difficult it will be for it to follow the movement imposed by the fluid. This is because the force of inertia opposes all the changes related to the particle's movements. When a particle comes across a bubble, inertial impaction favors the deposition of the particle on the bubble's surface by preventing a change in the movement direction. Centrifugal forces and bubble deformation are also involved in this mechanism before the attachment stage.

This liquid film is drained afterwards, as the particle glides along the bubble's surface. This is known as the adhesion or attachment stage. Not all particles coming across a bubble become attached to its surface. After the collision, there is still a thin layer of liquid that separates them. The thickness of this liquid must be drained as the particle slides along the surface. After the rupture of the liquid layer, a triple contact line between the liquid, the air bubble and the solid particle is formed.

Finally, the liquid film reaches its critical thinness and ruptures, enabling the bubble and the particle to create a stable bond that neither the particle's inertia nor the liquid's agitation can break before the bubble–particle coupling reaches the free surface. This is the stability stage.

Due to gravity, the particle moves away from the streamline and can pose itself on the bubble due to sedimentation. This effect contributes to bubble–particle collision.

When the Reynolds number is low, it is the sedimentation rate that is used to characterize the effect of gravity. As the particle is attached to the bubble, their bond must resist external stress so that the aggregate can reach the flotation unit's surface without being destroyed. The stability of particle–bubble aggregates is generally limited by the effect of turbulence.

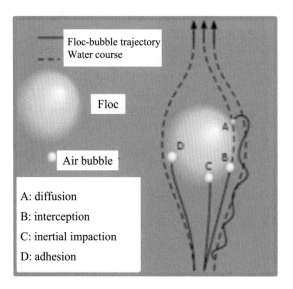

Figure 6.3. *Functional diagram of floc–bubble adhesion. For a color version of this figure, see www.iste.co.uk/gaid/watertreatment1.zip*

These three capture stages are controlled by the various interactions between the bubble and the particle: the hydrodynamic interaction, the gravitational interaction, the capillary interaction and the interaction due to surface forces. This can be summarized in a matrix in such a way that:

Probability of optimal flotation = (probability of floc-bubble collision) × (attachment probability) × (probability of non-detachment)

In flotation systems, the electric charges carried by the particles and the bubbles are of the same sign, inducing a repulsive electrostatic force.

6.3.1. *Coagulation*

Without the addition of coagulant or when its dosage is insufficient, air bubbles and particles carry negative charges. When particles approach the air bubbles, the electric double layers surrounding the particles and the bubbles overlap, provoking a repulsive force.

van der Waals forces between particles and air bubbles can be attractive (by providing a binding mechanism) or can be repulsive. The addition of a coagulant neutralizes particle charges in such a way that electrostatic forces become weak or close to zero. Particles become more hydrophobic, and air bubbles can adhere to them.

Briefly stated, coagulation is essential to reduce the repulsive charge interactions between particles/flocs and bubbles.

6.3.2. *Flocculation*

The purpose of flocculation is to increase the particle size (dp) for the attachment with the bubbles in the contact zone to become effective. As the particle or floc size increases, the efficiency of the contact zone improves due to physical interception mechanisms, such as sedimentation. The floc density is generally between 1,050 and 1,100 $kg \cdot m^{-3}$.

Suitable conditions for particle–bubble collisions require the floc size to be at least 100 μm. In this case, the efficiency of the contact zone approaches 100%.

Even if efficiency is >99% for 25–50 μm sized flocs, it is sufficient. This explains why the addition of a polymer in the flocculation tank must be examined on a case-by-case basis, depending on the quality of raw water.

In addition, the formation of large flocs can generate a bubble-detachment dynamic or require several bubbles to sufficiently reduce their density and reach high rising velocities.

6.3.3. *Contact zone*

6.3.3.1. *Bubble formation: gas solubility and stokes drift velocity*

Gas solubility in fresh water mainly depends on temperature, the type of gas and gas partial pressure. In addition to these three parameters, the solubility of gas in water also depends on salinity. According to Henry's law, the amount of air dissolved in saturated water depends on temperature and on the air pressure above the liquid. Henry's law is used for calculating the dissolved gas mole fraction of any gas, xi, at equilibrium conditions:

$$xi = Pi\ (h)/Hi$$

where:

– xi: molar fraction of air dissolved in water;

– Pi: partial pressure of the gaseous phase above the liquid (in this case, water) at height h (h = 1.0 if height 1000 m);

– Hi: Henry's law constant, which has the dimensions of pressure and is related to the nature of the dissolved gas (in this case, air). It depends on increasing temperature T when the temperature rises. The mole fraction of dissolved air decreases when the temperature rises.

Gas partial pressure is corrected depending on altitude:

$$Pi(h) = Pi(o)\ e^{-0,121h}$$

where:

– h: altitude above sea level in meters;

– Pi(o): gas partial pressure, i, at sea level where h is 0 m. The partial pressure of O_2 in the atmosphere is 0.2095 atm.

The dissolved gas concentration ($mol\cdot L^{-1}$) is calculated using the following approximation:

$$Ci = Xi\ Cwater$$

where:

– Ci: dissolved gas concentration, $mol\cdot L^{-1}$;

– Cwater: molar concentration of water:

$$Cwater = 1000\ (g\cdot L^{-1})/18\ (g\cdot mol^{-1}) = 55.56\ mol\cdot L^{-1}$$

For example, the mole fraction of O_2 in water at atmospheric pressure and at 20°C is:

$$X_{O2} = P_{O2}/H_{O2}$$

That is:

$$X_{O2} = 0.2095 \ (atm)/(4.09.10^4) \ (atm \cdot mol^{-1}) = 5.12 \times 10^{-6}$$

The dissolved oxygen saturation concentration (20°C) is then:

$$C_{O_2} = 5.12 \times 10^{-6} \times 55.56 \ (mol \cdot L^{-1}) \times 32,000 \ (mg \cdot L^{-1}) = 9.11 \ mg \cdot L^{-1}$$

The dissolved air concentration is the sum of the concentrations of the respective saturations in O_2 and N_2.

That is:

$$CS, air = CS, O + CS, N$$

The N_2 saturation concentration was calculated using the same method as for O_2. The mole fraction of N_2 in water at atmospheric pressure and at 20°C is:

$$X_{N2} = P_{N2}/H_{N2}$$

That is:

$$X_{N2} = 0.7809 \ (atm)/(8.04 \times 10^4) \ (atm \cdot mol^{-1}) = 9.71 \times 10^{-6} \ mol \cdot L^{-1}$$

Temperature (°C)	Henry's constant, O_2 atm·mol^{-1}	Henry's constant, N_2 atm·mol^{-1}	O_2 saturation concentration values, mg·L^{-1}	N_2 saturation concentration values, mg·L^{-1}	Air saturation concentration values, mg·L^{-1}
5	29,100	60,700	12.8	20.0	32.8
10	32,900	66,800	11.3	18.1	29.4
15	36,900	72,700	10.1	16.5	26.6
20	40,900	80,400	9.1	15.1	24.2
25	44,900	86,600	8.3	13.9	22.2
30	48,900	92,400	7.6	13.1	20.7

Table 6.1. *Henry's constant and saturation constant values of O_2 in water (h = 0 m sea level, salinity: 0 g·L^{-1})*

The dissolved N_2 saturation concentration (20°C) is then given as:

$$C_{N2} = 9.71 \times 10^{-6} \times 55.56 \ (mol \cdot L^{-1}) \times 28,000 \ (mg \cdot L^{-1}) = 15.1 \ mg \cdot L^{-1}$$

The dissolved air concentration at 20°C is:

$$CS, air = 9.1 + 15.1 = 24.2 \ mg \cdot L^{-1}$$

Temperature (°C)	10	20	30	40
Henry's constant atm·mol^{-1}	5.49×10^4	6.64×10^4	7.71×10^4	8.70×10^4

Table 6.2. *Henry's constant values for air as a function of temperature*

When a gas is in contact with a liquid's surface, the amount of gas in solution is proportional to the partial pressure of this gas. A simple justification for Henry's Law is that if the partial pressure of a gas is twice as high, on average twice as many molecules will hit the liquid's surface in a given time interval, and on average twice as many will be captured and go into solution. For a gas mixture, Henry's law is useful for predicting how much of each gas will go into solution. The parameters in Tables 6.1 and 6.2 are used for the calculations.

The flotation of particles associated with bubbles under laminar flow conditions is given by Stokes' law. It is used for calculating the rising velocity of bubbles in water. For a nearly massless gas bubble, the maximum velocity is reached when the buoyant force equals the drag force:

$$\pi d^3 (\rho w - \rho b)g = 18 \ \pi dv \ \rho w \ Vb$$

$$Vb = \frac{d^2 (\rho w - \rho b)g}{18 \ v \ \rho w}$$

where:

– db: diameter of a spherical gas bubble (m);

– Vb: rising velocity of a spherical bubble with a diameter db (m·s^{-1});

– ρw: water density (kg·m^{-3});

– ρb: density of the gas bubble, in (kg·m^{-3});

– v: kinematic viscosity of water (Pa·s);

– g: acceleration due to gravity (9.8 m·s^{-2}).

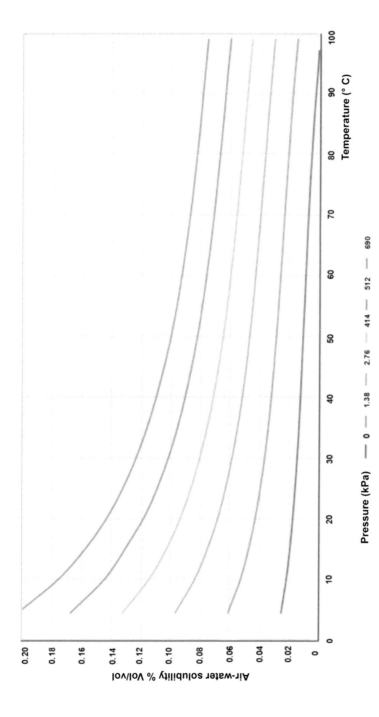

Figure 6.4. *Air solubility values in water (%v/v) at different pressures and temperatures. For a color version of this figure, see www.iste.co.uk/gaid/watertreatment1.zip*

For example, the rising velocity of a bubble with a 50 µm diameter at a 20°C water temperature is:

$$Vb = \frac{(50 \times 10^{-6})^2 (998 - 1.20)\, 9.81}{18 \times 998 \times 1{,}005 \times 10^{-6}} = 0.00135 \text{ m·s}^{-1} = 4.86 \text{ m·h}^{-1}$$

where:

- Vb: terminal velocity of the spherical bubble db (m·s^{-1});
- db: diameter of a spherical gas bubble (50×10^{-6} m);
- ρw: water density (998.2 kg·m^{-3});
- ρb: density of the gas bubble (1.20 kg·m^{-3});
- v: kinematic viscosity of water (1.005×10^{-6} m²·s^{-1});
- g: gravitational acceleration (9.81 m·s^{-2}).

Figure 6.5 shows the production of whitewater (photo on the left) from depressurized air bubbles within an injection nozzle (photo on the right).

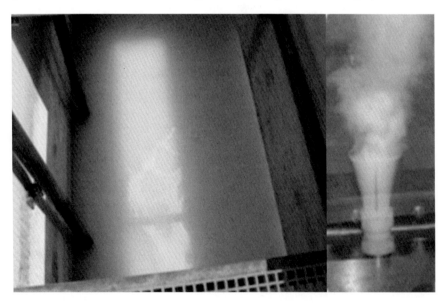

Figure 6.5. *Whitewater production from very fine air bubbles. For a color version of this figure, see www.iste.co.uk/gaid/watertreatment1.zip*

Figure 6.6 shows the variation in the rising velocity of air bubbles as a function of their diameter and temperature. At 20°C, 50 μm bubbles have a rising velocity < 5 m·h⁻¹, whereas 150 μm diameter bubbles have a velocity 10 times higher. Thus, in a reactor, the duration of air bubble immersion in water is greater the lower the bubble rising velocity (and therefore, their diameter). Without the use of coagulants, the lowest size limit for applying separation by flotation is approximately 35 μm. On the other hand, particles as small as 1 μm are separated in coagulated and flocculated water.

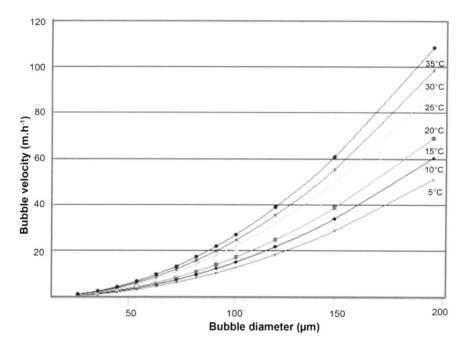

Figure 6.6. *Rising velocity as a function of bubble diameter at different temperatures (°C). For a color version of this figure, see www.iste.co.uk/gaid/watertreatment1.zip*

The dissolved oxygen concentration as a function of salinity and temperature is given by the ratio:

$$C_{O2} = \frac{(475 - 2.65\,S)}{(33.5 + T)}$$

where:

– C_{O2}: dissolved oxygen concentration at equilibrium with air above water (mg·L⁻¹);

– T: temperature (°C);

– S: dissolved salt concentration ($g \cdot L^{-1}$).

A key feature of DAF concerns the production of microbubbles, whose sizes may vary from a few μm to a few tens of μm. The energy requirements to generate air bubbles from pressurized recirculated water must be considered when designing DAF systems.

A properly functioning DAF implies that microbubbles have a variable size between 40 and 80 μm to obtain a suitable generation of aggregates in the contact zones and a proper separation in the separation zone. Bubbles only affect flotation insofar as they manage to adhere to the particles. This generally assumes that their diameter is smaller than that of particles or flocs present in the water arriving in the contact zone.

The water velocity when passing from the contact zone to the separation zone is in the range of 150–200 $m \cdot s^{-1}$, and bubbles larger than 500 μm rise directly to the contact zone's surface. The rising velocity of smaller bubbles is so low that they follow the water flow in the separation zone and contribute to the emergence of the low-density whitewater layer. In this way, stratification is obtained in the volume of the separation zone with a mainly horizontal water movement, showing a weak descending velocity in this layer.

The lower water layer is essentially characterized by a homogeneous vertical flow rate given by the ratio between the water flow and the separation zone's mirror surface.

Slowly rising bubbles into the upper layer collide against a steadily descending floc, and larger aggregates can form up in this zone. Bubbles that are not involved in the formation of aggregates become stuck in the surface sludge layer, making it more stable.

Bubbles with a rising velocity lower than the flow velocity are carried by the water's downward movement. That is, they move toward the outlet of the water produced. This bubble bed can be voluminous and degrade the quality of floated water because these bubbles can push fine particles.

A balance of forces acting on a microbubble placed in the descending flow of the separation zone makes it possible to calculate the bubble's drag limit, depending on its size and flow velocity. The rising velocity of an air bubble responds to two opposing forces. First, the differential densities of air and water generate a net upward buoyancy force. Second, the bubble encounters a drag force that resists upward motion.

In terms of force balance, we can state that the buoyancy force and the drag force cancel each other out.

The resulting force balance is expressed as

Buoyancy Force – Drag Force = 0

A diagram presents these aspects in Figure 6.7 for a bubble placed in a descending flow, that is, at the flotation unit's separation zone.

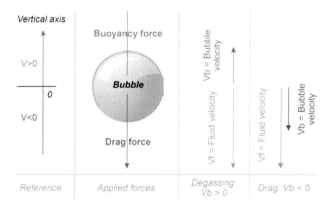

Figure 6.7. *Diagram representing the forces and velocities of a bubble in a descending flow. For a color version of this figure, see www.iste.co.uk/gaid/watertreatment1.zip*

The equation of the force balance is as follows:

$$\frac{4}{3}\pi R^3 (\rho_b - \rho_f)\vec{g} - Cd\frac{\pi R^2}{2}\rho_f \left\|\overrightarrow{V_{bf}}\right\|\overrightarrow{V_{bf}} = \vec{0} \text{ where } \overrightarrow{V_{bf}} = \overrightarrow{V_f} - \overrightarrow{V_b}$$

where:

– R: floc radius (m);

– g: acceleration due to gravity (m·s^{-2});

– Cd: drag coefficient;

– Vbf: floc–bubble aggregate velocity (m·s^{-1});

– Vb: bubble velocity (m·s^{-1});

– Vf: floc velocity (m·s^{-1});

– ρb: bubble density (kg·m^{-3});

– ρf: floc density (kg·m^{-3}).

By projecting the vector equation along the vertical axis:

$$V_{bf} = \frac{4}{3}\frac{d(\rho_b - \rho_f)g}{Cd\rho_f V_{bf}}$$

The Reynolds number is a dimensionless number defined as:

$$Re_b = \frac{d\|\overrightarrow{V_{bf}}\|\rho_f}{\mu_f}$$

The relative velocity can then be expressed as a function of the Reynolds number:

$$V_{bf} = \frac{4}{3}\frac{d^2(\rho_b - \rho_f)g}{Cd\,Re_b\,\mu_f}$$

where CD is the drag coefficient valid for isolated spherical solid particles, as well as for bubbles with a contaminated interface

$$Cd = \frac{24}{Re_b}\left(1 + 0.15Re_b^{0.687}\right) \text{ for } Re_b \leq 1,000$$

$CD = 0.44$ for $Re_b > 1,000$

The tests carried out by VERI-Veolia by iterative resolution made it possible to determine the bubble velocity depending on its size as well as the fluid's velocity (carried out using MATLAB). Water velocity values are set from 0 to -70 m·h^{-1} (negative vertical velocity, due to the descending flow), and the bodies present are water and air at a temperature condition of 20°C. The results obtained for the bubble velocity are shown in Figure 6.8, representing a response surface as a function of water velocity and bubble size. While the red color indicates positive bubble velocities in the reference (degassing bubbles), the blue color denotes negative bubble velocities, that is, those dragged by the flow. Solid black lines represent bubble iso-velocity curves, whereas dotted black lines display iso-diameters.

These results indicate an air bubble entrainment lower than 10 µm, regardless of the liquid's velocity values. Beyond this size, there is a drag limit that corresponds to the bubble's zero velocity. This threshold corresponds to the dividing line between the colors red and blue in Figure 6.8. This drag limit curve is shown in Figure 6.9. Then, according to the previously stated theoretical hypotheses, it would appear that a 50 µm diameter bubble remains motionless in a 4.7 m·h^{-1} downward vertical flow. For high-velocity flotation units designed for a 30–45 m·h^{-1} vertical flow rate in the separation zone, the calculation predicts a 141 μm drag limit. Theoretically, to be able to degas, microbubbles should have at least this size limit for real-life applications.

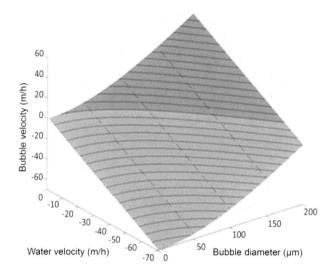

Figure 6.8. *Air bubble velocity in a flow. For a color version of this figure, see www.iste.co.uk/gaid/watertreatment1.zip*

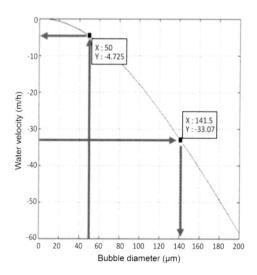

Figure 6.9. *Drag limit or descending water degassing, depending on a bubble's fluid velocity as well as on its size (red Vb > 0/blue Vb < 0) and the descending water velocity. For a color version of this figure, see www.iste.co.uk/gaid/watertreatment1.zip*

This calculation shows the risk of air bubble entrainment for 50 μm microbubbles, unless these microbubbles are incorporated into a floc aggregate.

In practice, when air-saturated water is gradually released, a minimum pressure difference is required to generate bubble nucleation. A large pressure difference across the air injection nozzle produces bubble nuclei. Assuming that air is an ideal gas, the critical bubble nucleus diameter (db) for homogeneous nucleation is the result of the following equation:

$$db = \frac{4\Upsilon}{\Delta P}$$

where:

– Υ: surface tension (72.8×10^{-3} N·m^{-1} at 20°C);

– ΔP: pressure difference across the nozzle (bar).

Figure 6.10 shows the evolution of bubble diameter as a function of pressure variation.

In DAF, bubbles are formed by reducing the pressure of a water stream previously saturated with air until reaching a pressure value above 3 bar. In industrial practice, supersaturated water is forced through specific nozzles producing bubble clouds with a diameter between 20 and 100 μm.

The air saturation concentration (mg·L^{-1}) at the outlet of the gas saturation tank, and depending on pressure and temperature, is:

Cr (air) = $sT \times P \times 12.96$

Pressure P is expressed in bar, and sT is the solubility of the gas in water at temperature T°C. Air saturation concentrations can be calculated for different pressure and temperature values (Table 6.2). For example, for a pressure of 5 bar and an air solubility at 20°C equal to 1.87 (% v/v), the saturation concentration will be

Cr(air): $1.87 \times 5 \times 12.96 = 121.2$ mg·L^{-1}

The bubble suspension (whitewater blanket) in the contact zone reaches a steady bubble concentration as a result of the continuous inflow of the pressurized recycled stream (Qr), which mixes with the influent water (Q) coming from the flocculation tank.

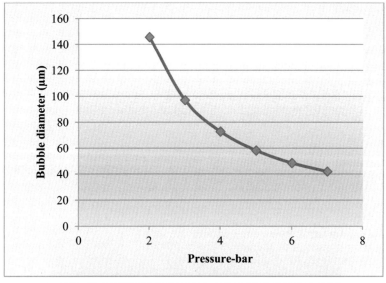

Figure 6.10. *Bubble diameter depending on pressure (graph) and the nozzle installation device (photo). For a color version of this figure, see www.iste.co.uk/gaid/watertreatment1.zip*

Three types of measurements characterize bubble concentrations: mass, volume, and number.

The oxygen concentration Cb (mg·L^{-1}) in the contact zone is:

$$Cb = \frac{e(Cr - CS)R}{(1 + R)}$$

where:

– Cb: water O_2 concentration in the contact zone (mg·L^{-1});

– e: saturator efficiency (90%);

– Cr: recirculated water O_2 concentration at T°C (mg·L^{-1});

– CS: inlet water's saturation concentration (mg·L^{-1});

– R: recirculation rate.

e is a transfer coefficient (0.9).

To find the volume concentration of bubbles (ϕb), Cr is divided by the saturated air density (ρa):

$$\phi b = \frac{Cb}{\rho a} = \frac{e(Cr - CS)R}{\rho a\,(1 - R)}$$

where ρa is 1.204 kg·m^{-3} at 20°C.

6.3.3.2. *Evaluation of the number of air bubbles*

The diameter of the air bubbles (db) gives access to the number of air bubbles (Nb). The bubble sizes formed mainly depend on the saturator pressure and the injection device (nozzle type or needle valve) at 4–8 bar saturator pressure. Bubble dimensions are between 20 and 150 µm, with most of the bubbles approximately 40–80 µm, with an average size of 50 µm. The density of air bubbles is equal to the air/water ratio divided by the volume of an air bubble.

That is:

$$Nb = \frac{\phi b}{(\pi db^3)/6} = \frac{e(Cr - CS)R}{\rho a\,(1 - R)((\pi db^3)/6)}$$

The results shown in Table 6.3 are obtained at 20°C.

P (bar)	db (μm)	Cr (mg·L⁻¹)	Cb (mg·L⁻¹)	φb ppm	Number of bubbles/mL	Number of bubbles/mL
					$R = 10\%$ at 20°C	
2	145.6	48.5	1.91	1,605	909	908,729
3	97	72.7	3.80	3,193	6,102	6,101,804
4	72.8	96.9	5.65	4,748	21,505	21,504,993
5	58.2	121.2	7.50	6,303	55,755	55,754,786
6	48.5	145.4	9.40	7,899	120,751	120,751,485
7	42	169.6	11.20	9,412	226,197	226,197,238
					$R = 5\%$ at 20°C	
2	145.6	48.5	1.00	840	476	808,816
3	97	72.7	1.98	1,664	3,179	475,774
4	72.8	96.9	2.96	2,487	11,266	3,179,361
5	58.2	121.2	3.94	3,311	29,290	11,266,332
6	48.5	145.4	4.92	4,134	63,202	29,289,848
7	42	169.6	5.90	4,958	119,157	63,201,841

Table 6.3. *Number of bubbles/L depending on pressure and bubble diameter*

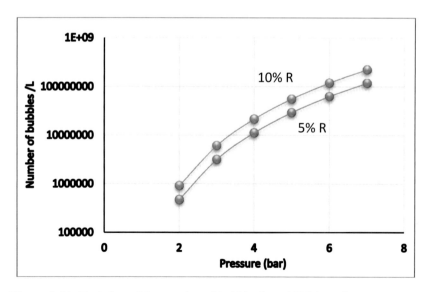

Figure 6.11. *Evolution of the number of bubbles/L at 20°C for different pressure values and two recirculation rates. For a color version of this figure, see www.iste.co.uk/gaid/watertreatment1.zip*

6.3.3.3. *Bubble size evaluation*

The bubble size depends on various parameters:

– physicochemical characteristics of the water used for the preparation of whitewater;

– operating conditions of a gas saturation tank: pressure, percent air saturation at working pressure;

– pressurized water relief conditions.

Bubble size is affected by water pressure, viscosity and surface tension. At sea level, the specific weight of air varies from 1.24 g·L^{-1} at 10°C up to 1.16 g·L^{-1} at 30°C. A standard atmospheric pressure of 10.333 m of water = 101.325 kPa = 1.013 bar.

Bubbles are formed by cavitation due to the pressure drop in the injection nozzle. First, they form their nuclei, and then they start to develop. The bubble size depends on the saturator pressure at the outlet and the injection flow rate.

High flow rates produce smaller bubbles, and the bubble diameter becomes constant at the maximum flow rate in relation to nozzle dimensions. This is because the nozzle geometry has a strong influence on the bubble size distribution. This is related to the degree of turbulence caused by the passage of the liquid stream through the nozzle: an increase in the mass transfer from liquid to gas results in a smaller or higher number of bubbles. The longer the nozzle is, the larger the bubbles formed.

Small bubbles (sizes between 20 and 150 µm) in the form of rigid spheres are most often produced by DAF systems. This is an important parameter since it affects collision performance, particle attachment to bubbles and their rising velocity.

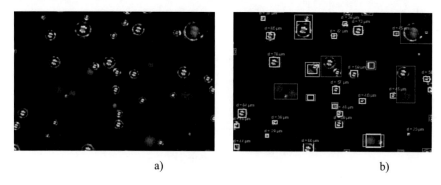

a)   b)

Figure 6.12. *Examples of microbubble raw images (a) and of their MATLAB analysis (b). For a color version of this figure, see www.iste.co.uk/gaid/watertreatment1.zip*

Nevertheless, the actual bubble size in DAF systems is affected by heterogeneous nucleation, bubble growth, the injection flow rate, and the injection device. The saturator pressure and the injection device, and in particular the type of nozzle, are important factors that influence the bubble size.

The removal efficiency increases with floc size, but after a certain point, efficiency becomes constant. As the volume concentration of bubbles increases, the collision between particles and bubbles increases concomitantly, which results in a better attachment efficiency.

The higher the pressure is, the smaller the bubble size. For example, Table 6.3 shows the size distribution of bubbles at different pressure values. The lower the pressure is, the wider the bubble size distribution, and the smaller the bubble number ratio. The ratio between average floc size, pressure and efficiency shows the effect of pressure on efficiency. The critical floc size at which efficiency does not change is 32 μm for 5 bar and 40 μm for 3 bar.

Tests carried out by VERI-Veolia show that in the contact zone, the majority of bubbles formed have a diameter of approximately 30 μm. This population also evolves from the bottom of the contact zone toward the surface, where bubble enlargement is observed. Several phenomena may explain this:

– a modification in gas density by lowering the pressure of the water column above;

– a coalescence of microbubbles among themselves due to particle shocks;

– a diffusion of dissolved air extending into the mixing zone and reaching the bubbles already formed.

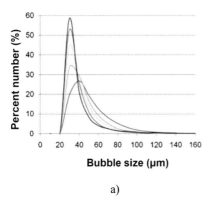
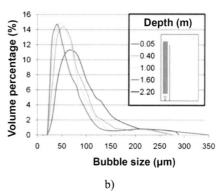

a)   b)

Figure 6.13. *Size distributions of bubbles according to their number (a) and volume (b) in the contact zone. For a color version of this figure, see www.iste.co.uk/gaid/watertreatment1.zip*

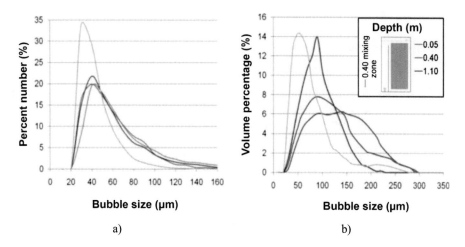

Figure 6.14. *Size distributions of bubbles according to their number (a) and volume (b) in the separation zone. For a color version of this figure, see www.iste.co.uk/gaid/watertreatment1.zip*

In the separation zone, the bubble sizes found are larger, with a d50 volume at the top of the zone equal to 140 µm. A segregation in terms of size and height also appears in that zone. Since fine bubbles no longer follow the flow, the size distribution shifts toward smaller sizes as they approach the separation zone.

	Diameters	Number	Volume
Top of the contact zone	Modal d (µm)	40 µm	70 µm
	d50 (µm)	50 µm	80 µm
Top of the separation zone	Modal d (µm)	40 µm	100–140 µm
	d50 (µm)	50 µm	140 µm

Table 6.4. *Characteristic diameters of microbubbles at the top of the contact and separation zones (mirror velocity: 33 m·h⁻¹)*

This VERI-Veolia study revealed the existence of a median bubble size with a volume proportion of 80 µm in the mixing zone and 140 µm in the separation zone. The analysis by number still reveals a large amount of 50 µm diameter bubbles.

6.3.3.4. Flocs and agglomerates visualization

Photographs make it possible to qualify the flocs formed according to their size and geometric shape (Figures 6.15 and 6.16).

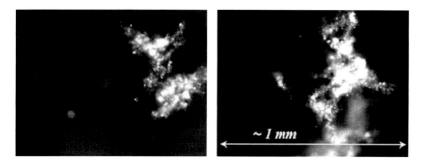

Figure 6.15. *Photographs of flocs*

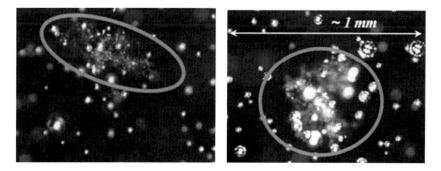

Figure 6.16. *Photographs of floc–bubble agglomerates*

A local analysis of the process revealed the impossibility of degassing isolated 50 μm bubbles at the flow rates used in fast DAF systems. According to Stokes theory, 50 μm bubbles follow fluid-like tracers, and this type of flow amounts to performing a single-phase simulation because the phases never actually separate. For example, the drag limit of a bubble subjected to a downward vertical flow of 33 m·h^{-1} is 141 μm in diameter. This size corresponds to the average size measured per volume at the top of the separation zone.

The population of microbubbles evolves within the structure. Even if the usually mentioned DAF of 50 μm still exists, this size no longer represents the majority per volume in the separation zone. Floc visualization shows an attachment of several bubbles per floc.

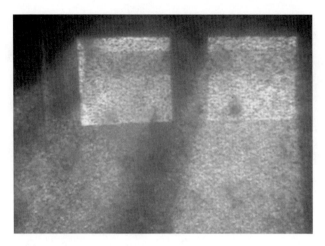

Figure 6.17. *"Flock-bubble" coupling. For a color version of this figure, see www.iste.co.uk/gaid/watertreatment1.zip*

6.3.4. *Separation zone*

The removal of air bubble–floc aggregates takes place in the separation zone. The design of the separation zone and the removal performance of free bubbles and floc–bubble aggregates are based on Hazen's theory, analogous to settling theory.

A vertical flow is present all along the main portion of the solid–liquid separation zone, which is actually the clarification zone. The free bubbles and floc–bubble aggregates are retained in the upper part of the separation zone if their rise velocities (vb and vfb) are higher than the water's descending velocity.

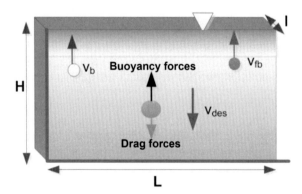

Figure 6.18. *Functional diagram of the solid–liquid separation. For a color version of this figure, see www.iste.co.uk/gaid/watertreatment1.zip*

Assuming a laminar flow and spherical aggregates, the aggregate rise velocity (vfb) can be calculated according to Stokes' law:

$$vfb = \frac{gdag^2(\rho w - \rho ag)}{18\,\eta\rho w}$$

where:

- ρag: aggregate density ($kg \cdot m^{-3}$);

- ρw: water density ($kg \cdot m^{-3}$);

- dag: aggregate diameter (m);

- η: kinematic viscosity of water ($1{,}005 \times 10^{-6}$ $m \cdot s^{-2}$ at 20°C);

- g: acceleration due to gravity ($m^2 \cdot s^{-1}$);

Since water density is far superior to that of the floating aggregates, this equation is often reduced to:

$$vfb = \frac{gdag^2}{18\,\eta}$$

In practice, efficiency is the result of:

$$R\% = 1 - \frac{Cout}{Cin}$$

where:

- Cin: inlet turbidity (or SS) concentration ($mg \cdot L^{-1}$);

- $Cout$: outlet turbidity (or SS) concentration ($mg \cdot L^{-1}$).

Given the fact that gravity separation processes only rely on the difference in density between suspended solids and water, the separation of light particles is more difficult.

The main advantage of DAF is that it widens the density difference by making the floc–bubble aggregate buoyant.

The density of the floc–air bubble aggregate is deduced from:

$$\rho ag = \frac{\rho f df^3 + Nb\rho bdb^3}{df^3 + Ndb^3}$$

where:

- ρfb: density of the floc–bubble aggregate (kg·m^{-3});

- ρf: floc density (kg·m^{-3});

- ρb: air bubble density (kg·m^{-3});

- df: floc diameter (m);

- db: mean bubble diameter (m);

- Nb: number of bubbles attached to flocs.

The value of the floc–bubble aggregate diameter (dfb in m) is given as:

$$dfb = (df^3 + Nbdb^3)^{\frac{1}{3}}$$

For 1 < Re < 50, we use a drag coefficient CD:

$$CD = \frac{45}{Re^{0,75}}$$

The ascent rate (rising velocity) of floc–bubble aggregates in m·h^{-1} is given as:

$$vafb = \left[\frac{g(\rho w - \rho fb)dfb^{1.75} + Nb\rho bdb^3}{33,5\, \rho w^{\ 0.25}\mu^{0.75}}\right]^{0.8}$$

where:

- g: acceleration due to gravity (m·s^{-2});

- ρw: water density (kg·m^{-3});

- μ: dynamic viscosity (kg·m^{-1}·s^{-1}).

The separation velocity (or "descending" velocity) of water varies depending on the nature of the suspensions to be treated, as well as on the microbubble generation and distribution modes.

In a DAF unit, the separation velocity and the concentration of floating sludge are strongly influenced by the ratio:

R = amount of dissolved air/amount of floating matter

The R ratio can change the amount of air actually trapped by the floc. The increase in R can increase the reactor's operating velocity and therefore improve the sludge concentration.

While the presence of very fine bubbles is favorable to their distribution over the entire surface and their separation efficiency, this can nonetheless limit the flow rate, and consequently, the production of treated water.

The problem comes down to obtaining a sufficient air volume/floc volume ratio, which is governed by three mechanisms:

– the surface's affinity with bubbles, depending on the nature of the suspended solids (degree of surface hydrophilicity);

– an effective bubble–floc mixture;

– the availability of bubbles of the proper size to encourage efficient adhesion, between 20 and 80 μm.

Total efficiency approaches the value 0.9–0.95, not 1. This may be due to the presence of a short circuit in the contact zone, leading to a decrease in efficiency, or to the limited influence of the air/aggregate ratio in the separation zone.

High-rate DAF systems have a hydraulic loading rate of 25–40 $m \cdot h^{-1}$. The descending velocity follows a plug flow and is influenced by sludge extraction turbulence, the length/width ratio, the air bubble bed and the bed of particles, which tend to be drawn downward.

However, a stratification of concentrated sludge is observed in the upper portion of the separation zone, which extends across the entire surface of the contact and separation zones. Not only do bubbles produce aggregates with particles, but they also have an important role in the stratification of accumulated surface sludge.

This concentrated sludge plays a buffer role in relation to transverse velocity and acts as a sludge bed while maintaining aggregates and free bubbles.

However, to prevent the bubble bed from being dragged by the high flow velocity and degrading the quality of the treated water (because it is associated with aggregates), DAF designers have proposed a variety of systems with a head loss to block the bubble bed. These systems include lamellar blocks, intermediate nets, anti-spiral flow plates, etc.

Figure 6.19. *Separation zone before (a) and after (b) reagent injections. For a color version of this figure, see www.iste.co.uk/gaid/watertreatment1.zip*

6.4. Design parameters

While in a settler, it is necessary to obtain large flocs by flocculation for quick and easy settlement, flotation does not require such voluminous flocs. A large floc requires the attachment of many bubbles to make it buoyant. Air bubbles find it easier to adhere to finer flocs. The contribution of the positive charges provided by the injected coagulant is sufficient to neutralize the particles' negative charges and to prompt flocculation, producing flocs of approximately 100–300 μm.

DAF performance depends on coagulation, flocculation and the implementation conditions of supersaturated water for the production of microbubbles.

In high-rate DAF systems, the flotation unit's mirror velocity can be between 25 and 40 m·h^{-1}, or even more, while its geometry can be adapted to match this hydraulic requirement. The direction of the water flow is vertical from the water's free surface to the flotation unit's bottom. This is why the flotation zone is a square or a rectangle. The depth of the separation zone must be large enough because the microbubbles risk being carried downward together with the water. For bubble diameters between 50 and 80 μm, the total depth of the separation zone should be between 3.5 and 5.0 m. This means that in some cases, when the flow is low, the flotation unit's length and width may be shorter than the height.

6.4.1. *Coagulation*

Good coagulation is therefore essential to obtain favorable particle–bubble fixation. Without coagulation, particles carry a negative charge and are often hydrophilic, entailing poor fixation to bubbles. Optimal coagulation conditions are related to hydraulic conditions, coagulant dose and pH.

The advantage of a flotation unit compared to a settler is that it requires lower dosages of coagulant without the systematic addition of a polymer.

On the other hand, in cold water (<5°C), the use of a polymer is recommended because flocculation is more sensitive to this temperature.

The use of iron salts or a polymerized aluminum coagulant improves the kinetics and efficiency of the coagulation process, making it less sensitive to the need for pH optimization.

Static mixers are being increasingly used as a replacement for tanks with high-rate impellers. However, coagulation tanks are still necessary when the variation in the quality of raw water requires more stringent operating conditions.

The value of the velocity gradient G is between 100 and 180 s^{-1}, with a variable Gt between 6,000 and 18,000. A mixing energy of 78 g (rotation velocity) is recommended for the coagulation tank. The contact time is often between 1 and 2 min. The impeller is designed with a rotational velocity of 150 rpm^{-1}.

6.4.2. *Flocculation*

Flocculation is also an important stage, even if the flocs do not need to be as large as the ones obtained for settling. As much as settling requires heavy flocs to facilitate sedimentation, flotation requires small flocs, so that these can be easily "transported" by the bubbles to the surface.

The recommendation is to associate a single flocculation tank with each flotation unit, rather than one tank feeding several flotation cells. This is to prevent hydraulic short circuits or the poor distribution of hydraulic loads in each contact zone of the flotation unit, which would ultimately result in a poor performance of the bubble–floc collision and attachment mechanisms.

The flocculation tank is built within one or two baffled cells especially designed to provide flow conditions that prevent short-circuiting phenomena.

Flocculation times may vary between 10 and 20 min, depending on water quality and temperature, because small resistant flocs of varying diameter between 10 and 100 μm can be properly separated by flotation.

The value of the velocity gradient G is between 40 and 80 s^{-1}, with a variable Gt between 24,000 and 96,000. The recommended mixing energy is 30–40 g (rotation velocity). The impeller is designed with a rotational velocity of 40–50 rpm^{-1}.

To optimize the efficiency of the flocculation stage, a patented Veolia device, Turbomix, is installed inside the flocculation zone to improve mixing and to increase the outflow. This device ensures an optimal use of all chemicals and a smaller flocculation tank footprint.

Turbomix ensures:

– a high level of homogenization and low retention time;

– the removal of dead zones;

– a thorough mixing of the inlet water with the chemical reagents;

– zero or very low polymer consumption.

By integrating Turbomix, contact times can be reduced to 6 min for temperatures of 20°C and above.

6.4.3. *Flotation zone*

Significant developments have been made in the design and operation of DAF units over the past 10 years. Regarding cell geometry, the square or rectangular cell is almost universally used, and this is due to:

– its simpler construction;

– the fact that multiple equivalent units are cheaper to build and therefore leave a smaller print;

– hydraulic distribution is easy to implement with especially adapted equipment.

The size of DAF units depends on the hydraulic loading rate, considering that mirror velocities vary between 25 and 40 m·h^{-1}. The solid–liquid separation process requires laminar flow conditions. As a result, the optimal arrangement of the contact zone's outlet (which is equipped with an inclined deflector) reduces the rise velocity and enhances the horizontal flow for a better stratification of sludge.

The outlet of floated water is often done through perforated side pipes. Problems caused by recirculating currents as well as descending bubble beds (whose depth may vary between 1.5 and 2 m) affect the quality of floated water. To solve them, it is necessary to set up a special type of equipment within this zone (lamellar pack, anti-spiral flow plates, intermediate sieve, etc.) to create a head loss and prevent the bubble bed from descending to the level of the floated water outlet.

This equipment makes it possible to stabilize the hydraulic effects in the separation zone, reduce turbulence and recirculation currents, encourage the separation of the air/solid phase in clarified water and prompt the degassing of excess air, the formation of larger bubbles and their rise to the surface. They contribute to the separation of the sludge flow from the descending water flow.

All these facilities now make it possible to offer larger flotation sizes, among which Veolia, for example, can supply 100 m² flotation units (Spidflow®). A unit of treatment based on an average operating mirror velocity of 30–40 m·h^{-1} can treat 3,000 m³·h^{-1} (72,000 m³·d^{-1}).

Figure 6.20. *Surface scraper. For a color version of this figure, see www.iste.co.uk/gaid/watertreatment1.zip*

Width (10 m maximum) is limited by the maximum reach of most mechanical scrapers. When the width is increased, sludge removal is achieved by increasing the water level rather than by using a scraper.

6.4.3.1. *Contact zone*

The flocculated water enters the contact zone by openings placed at the floor level. The rising velocity profile shows a decrease along the trajectory because the space tends to widen in the upper area. This is due to a greater width, as the wall separating the contact zone from the separation zone tends to widen. This happens even if the flow remains in plug flow mode. At the top of the contact zone, velocities become horizontal, transporting the floc–bubble aggregates toward the flotation unit's posterior wall. The contact time in this zone should be 90–120 s.

The distance between the water surface and the top of the separation wall (contact zone-separation zone) should not be too considerable, because the flow structure could be irregular, leading to a risk of bubble accumulation above the contact zone and increasing the probability of backflow into the contact zone. For this reason, it is important to have a high horizontal flow velocity from above the contact zone toward the separation zone. The separation between the contact zone and the separation zone is obtained by using a deflector.

All suspended solids in the contact zone must have a sufficient rise velocity v to cross the effective height H during residence time t:

$$vfb = \frac{H'}{t}$$

where:

– vfb: rise velocity ($m·h^{-1}$);

– H': effective depth of the contact zone (m);

– t: detention time (h).

The rise velocity in this zone is between 130 and 200 $m·h^{-1}$.

6.4.3.2. *Separation zone*

The design of the separation zone takes into account not only the mirror velocity but also the rise velocity of the floc–bubble aggregates. The mirror velocity is obtained by the ratio between the flow rate through and the flotation surface:

$$v \, mirror = \frac{Q}{S}$$

where:

– v mirror: mirror velocity ($m·h^{-1}$);

– Q: inlet flow ($m^3·h^{-1}$);

– S: mirror surface of the separation zone (m²), that is, the separation zone surface of the footprint area. For a common rectangular DAF tank, the surface of the separation zone is the length (L) times the width (W).

The particles to be removed must have a sedimentation rate lower than the rise velocity. In practice, variable mirror velocities between 15 and 40 m·h⁻¹ are applied.

In general, the design of a DAF unit involves not exceeding the maximum velocity recommended for the different zones (Figure 6.21). Velocity v_1 at the flocculation tank's outlet feeding the contact zone's feed-in channel must not be too high to prevent floc shear. The same applies to velocities v_2 and v_3, which are the flow velocity in this channel and the inlet velocity to the contact zone, respectively.

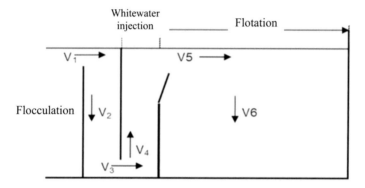

Figure 6.21. *Velocities involved in a flotation unit. For a color version of this figure, see www.iste.co.uk/gaid/watertreatment1.zip*

Velocity v_4 is the vertical velocity in the contact zone. It must be lower than the largest bubbles' rise velocity and at the highest temperature.

Velocity v_5 is the horizontal velocity in the flotation zone (or separation zone). If the velocity is too high, it drags the flocs too abruptly toward the opposite wall of the flotation unit and provokes a spiral flow that can cause the sludge to settle. Velocity v_6 is the downward water velocity relative to the water flow near the outlet.

The height H′ is set between 3.5 and 5 m to avoid excessively high flow gradients, and the rectangular configuration is preferred for large surfaces to operate in piston flow mode.

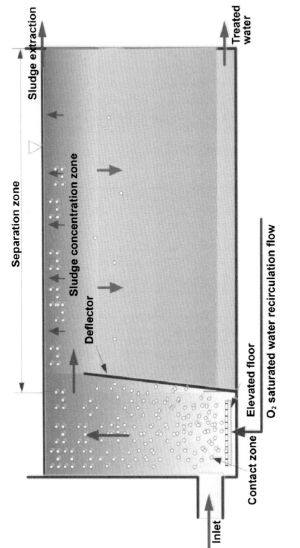

Figure 6.22. *Operating principle of a flotation unit. For a color version of this figure, see www.iste.co.uk/gaid/watertreatment1.zip*

A length-to-width ratio of approximately 1.8–2.0 is recommended for maximum bubble removal in the separation zone. A horizontal piston flow is obtained, reaching the flotation unit's front wall and then moving back to the deflector. When the ratio is equal to 1, bubble removal may be lower, and the formation of horizontal flow layers may be far from optimal (Figure 6.22).

The usual length values oscillate between 3 and 10 m, and the width values vary between 1.8 and 6 m. The hydraulic retention time is 8–15 min. Recirculation rates range from 5–10% on average to 20% when required (algal bloom).

Finally, the water flow rate must be uniformly distributed and collected over the separation zone's entire surface to reduce dead zones.

6.4.4. Contact time

Table 6.5 summarizes the different contact times throughout the various stages.

	Coagulation	Flocculation	Contact zone	Separation zone	Gas saturation tank
Contact time	1–2 min	10–20 min	80–180 s	8–15 min	80–130 s

Table 6.5. *Contact time in the different zones of a flotation unit*

6.4.5. Temperature

Water temperature affects water viscosity, simultaneously influencing coagulation, flocculation and separation. The rise velocity of an aggregate (floc–bubble) in the contact zone and in the separation zone is 1.6 times lower at 2°C than at 20°C. Therefore, the efficiency in the separation zone is lower at low temperatures.

6.4.6. Air saturation tank

The concentration of air in solution (mg·L^{-1}) is given by the ratio between air density (at T°C and 1 atm, which is the reference pressure) and Henry's constant:

$$Cair = \frac{\rho air}{H}$$

ρ air: density of air

The concentration of air in solution (at the operating T°C and P values) is then given as:

$$C\ (T,P) = Cair\ \frac{Pon\ site}{Preference}$$

Henry's law is given as:

$$P = x_{air}.\ H$$

where:

– P: relative pressure in atm;

– $x_{air:}$ molar fraction of air in mol of air/mol of mixture (air + water);

– H: Henry's constant in atm, which varies depending on the nature of the gas, temperature and salinity.

The amount of air dissolved in a volume of water increases proportionally with pressure (Figure 6.23). Saturation concentration is given as a function of pressure (atmospheric pressure at sea level is 101.325 kPa) and as a function of temperature.

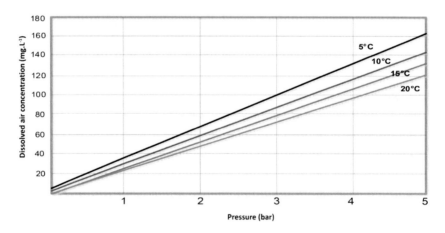

Figure 6.23. *Air solubility as a function of pressure at different temperatures. For a color version of this figure, see www.iste.co.uk/gaid/watertreatment1.zip*

6.4.6.1. *Design of the air saturation tank*

The water volume (*Vwater*) in the saturator is practically constant and at most must represent 75% of the total volume due to the control of the air injection by the water level sensor.

In stationary mode, when the water level increases (a sign of a decrease in internal pressure due to the consumption of air by dissolution), the sensor controls the air injection. The addition of air allows an increase in the pressure in the saturator and lowers the water level. Once the level has dropped below the sensor level, the injection is cut off. The cycle begins again. This operation mode guarantees constant whitewater quality.

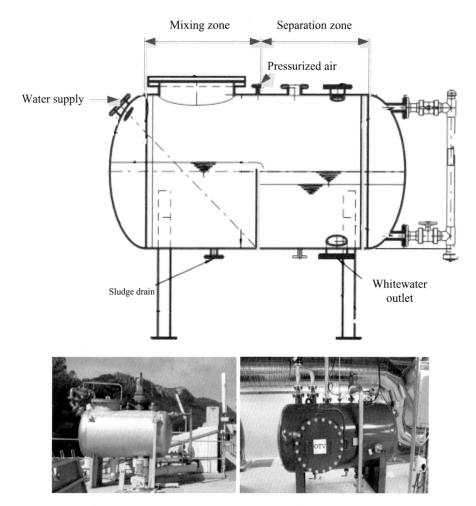

Figure 6.24. *Horizontal gas saturation tank. For a color version of this figure, see www.iste.co.uk/gaid/watertreatment1.zip*

Part of the clarified water is taken away from the floated water collector by the recirculation pump, passed through a prefilter and injected into the saturator's "agitated" compartment by means of one or more hydroejector(s). This specific equipment improves the air/water mixture due to a Venturi effect, which pulls air from the tank's vapor cloud and promotes dissolution. When the system starts up, the air demand is greater to "inflate" the saturator: the injected air is not only dissolved but also used to form the vapor cloud inside the saturator.

A stilling plate is added to divide the tank into an agitated zone and a quiet zone and to prevent the aspiration of large bubbles into the network downstream. It has a head loss in the range of 10 cm Ce.

The design rules of the whitewater production system are as follows.

The retention time (ts) is between 80 and 130 s to guarantee the dissolution of air in water. This makes it possible to determine the minimum liquid volume in the saturator:

Vwater = Q sat x ts

The hydroejectors (N_{hyd}) are calculated depending on the maximum unit flow (Q_{hyd}) in $m^3 \cdot h^{-1}$.

The horizontal-type saturator has a length-to-diameter ratio of at least 1.5.

The hydraulic power dissipated by the system is lower than 2,200 W per water m^3 in the saturator. Otherwise, the quality of the whitewater could be reduced. Power is obtained by:

$$Pb = \frac{\Delta H \rho g Qww}{Vwater}$$

where:

– P_b: power dissipated in the saturator in $W \cdot m^{-3}$;

– ρ: density of water in $kg \cdot m^{-3}$ (998.2 $kg \cdot m^{-3}$ at 20°C);

– g: gravitational constant is 9.81 $m \cdot s^{-2}$;

– ΔH: head loss by hydroejector (0.7 bar);

– Qww: whitewater flow per saturator in $m^3 \cdot s^{-1}$;

– $Vwater$: water volume in the saturator in m^3.

6.4.7. *Mass balance*

The flotation unit's operating mass balance is shown in Figure 6.25. The total amount of dry matter produced by the addition of reagents into raw water is:

$$SST = SSin + 0.07 * H_e + K * D_{coag} + xFe + yMn$$

where:

– SST: total amount of dry matter generated, $mg \cdot L^{-1}$;

– $SSin$: suspended solids in raw water, $mg \cdot L^{-1}$;

– $SSout$: suspended solids in floated water, $mg \cdot L^{-1}$;

– He: the color of the raw water in Hazen degrees, $mg \cdot L^{-1}$ PtCo;

– K: the coagulant's precipitation coefficient;

– D_{coag}: commercial product coagulant dose, $mg \cdot L^{-1}$ (8–10% aluminum sulfate or 40–42% ferric chloride, which are the most commonly used);

– Fe: the amount of precipitate obtained ($mg \cdot L^{-1}$), by iron oxidation of Fe^{2+} into Fe^{3+}, in the case of preoxidation: $xFe = 1.91 \times [Fe^{2+}]$ where $[Fe^{2+}]$ is the Fe^{2+} concentration in raw water, $mg \cdot L^{-1}$;

– Mn: the amount of manganese oxide MnO_2 obtained ($mg \cdot L^{-1}$) by oxidation of Mn^{2+} into Mn^{4+} due to the action of $KMnO_4$, chlorine dioxide or ozone:

$$yMn = 1.58 * [Mn^{2+}]$$

The sludge flow to be extracted is given by:

$$Qfls = Qw \frac{SST - SSin}{Sfls - SSout}$$

where:

– SST corresponds to the particle concentration sum in raw water (together with the floc concentration after coagulant), as well as to the particles formed during a possible chemical oxidation of iron and manganese. If powdered activated carbon (PAC) is injected into the flotation unit, this extra dose also has to be counted for the calculation of SST;

– $Qfls$: floating sludge flow, $m^3 \cdot h^{-1}$;

– Qw: raw water flow, $m^3 \cdot h^{-1}$;

– $SSin$: suspended solids in raw water, $mg \cdot L^{-1}$;

– *SSout*: residual suspended solids in floated water, mg·L^{-1};

– *Sfls*: floating sludge concentration, mg·L^{-1}.

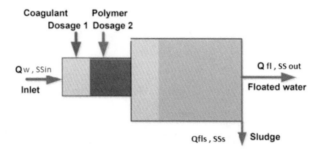

Figure 6.25. *Mass balance of a flotation unit. For a color version of this figure, see www.iste.co.uk/gaid/watertreatment1.zip*

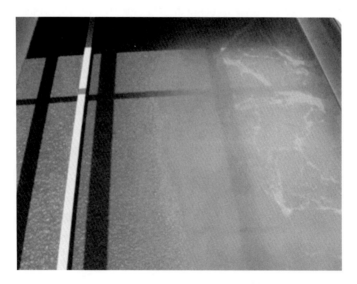

Figure 6.26. *Sludge concentration at the separation zone. For a color version of this figure, see www.iste.co.uk/gaid/watertreatment1.zip*

6.4.8. *Recirculation and injection nozzles*

The majority of energy requirements are related to the preparation of whitewater and its recirculation at a pressure level between 3.5 and 10 bar. Packed saturators are more efficient than empty or unpacked saturators (due to better regulation of O$_2$

transfer). However, given that saturators imply regular inspections at the operational level, they are unpacked saturators that are most often used. To this end, prefilters are installed at the saturator outlet to filter pressurized water and to protect the injection nozzles.

Most unpacked saturators are equipped with (internal or external) hydroejectors to dissolve pressurized air into water. The air that is separated within the saturator is recycled via the hydroejector.

Recirculation improves the efficiency of these saturators, even if they operate at a pressure 1–2 bar higher than packed saturators, for equivalent amounts of air.

The minimum operating pressure is approximately 3.5 bar. This is the reason why saturators generally operate between 4 and 6 bar. Higher pressures may be required when the solids load to be floated is high.

Their overall efficiency depends on the injection mode supplied. Injection nozzles are designed either with one or more fixed orifices or with an adjustable orifice such as the needle nozzle. The first system is the most widely used, although it is more sensitive to deposits and debris from supersaturated water. A prefilter placed upstream of the supersaturated water inlet can solve this drawback.

In practice, the quality of floated water is maintained with a recycling rate between 5 and 10% (with the possibility of increasing to 20%) and at pressure values between 4 and 6 bar, on average (with the possibility of rising to 8–10 bar). This is to produce a bubble density equivalent to an air requirement of 5–10 gm^{-3} for water at the indicated temperature. A good quality of water to be treated reduces energy consumption by modifying the recycling rate and the operating pressure.

6.5. Operating parameters affecting flotation performance

6.5.1. *Choice of coagulant*

The critical factor in the flotation process is floc rupture. When ferric chloride ($FeCl_3$) or aluminum sulfate ($Al_2(SO_4)_3$) is used as a coagulant, a floating water turbidity between 0.9 and 2 NTU is often obtained. This can be explained by a floc rupture in the contact zone, as these flocs are quite friable and can be affected by the turbulence of air bubbles. On the other hand, the use of a polymerized coagulant produces more resistant flocs, and the turbidity of the floated water obtained is better.

In general, the measurement of in-line turbidity is influenced by residual O_2 fine bubbles, which interfere with the measurement because the values shown are higher. The turbidity values obtained at the laboratory (degassing of the quiescent sample) and those recorded in-line are different, with higher values for continuous measurements.

Four parameters affect the design and functioning of the contact zone: bubble size, bubble rise velocity, contact time, and bubble volume concentration.

The collision efficiency between flocs and bubbles depends on bubble size. Small bubbles are better suited for flotation, which is why DAF is a more efficient process than dispersed air flotation, where the bubble size is much larger (0.7–1.2 mm).

As previously explained, the bubbles in the contact zone mainly have a diameter between 40 and 80 µm, with an average bubble size of approximately 50 µm. While it is true that small bubbles improve performance in the same way that the fine particle size of filters improves filtration, bubble size is primarily determined by the pressure difference across the whitewater injection device. It is therefore important for the design of the saturator and of the whitewater injection means to be properly developed. This is because the operator has little margin of control over bubble size, other than by maintaining the saturator's pressure.

A quick estimate shows that a 1 NTU turbidity level corresponds to 5,000 particles/mL (>2 µm). According to the previous calculation, we obtain a number of bubbles Nb equal to 37,468,259 bubbles·L^{-1} under the following conditions: 5 bar pressure, 20°C, 5% recirculation rate, 3 NTU turbidity, that is, a number of bubbles/particle ratios:

$$37{,}468{,}259/(3 \times 5{,}000{,}000) \neq 2.5\text{–}3 \text{ bubbles/particle}$$

However, to reach a turbidity of 15 NTU, the number of bubbles obtained at 5 bar is not sufficient. It is therefore necessary to work with a 10% recirculation rate and a 6 bar pressure, which makes it possible to obtain a number of bubbles equal to 149,195,351, that is, $149{,}195{,}351/(15 \times 5{,}000{,}000) \neq 2$ bubbles/particle. With a 7 bar pressure, we obtain 3–4 bubbles/particle, which is safer and more effective. It is therefore advisable to properly respect the recirculation rate and the saturation pressure to determine the best conditions for effective flotation. Even if this turbidity-to-particle number ratio is not precise, it makes it possible to estimate the conditions to be implemented at the level of the saturator and the recirculation rates.

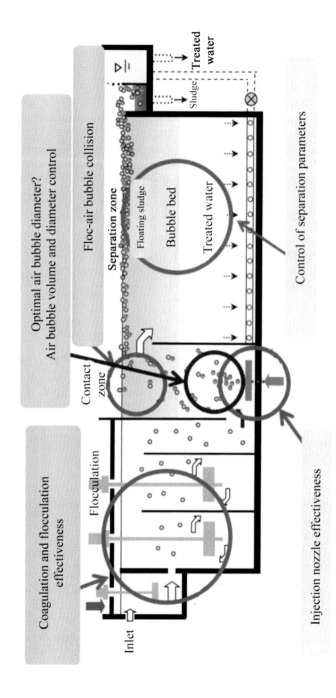

Figure 6.27. *Parameters influencing the flotation process. For a color version of this figure, see www.iste.co.uk/gaid/watertreatment1.zip*

6.5.2. *Rise velocity*

The rise velocity of bubbles can affect the performance of the contact zone, over which the operator has absolutely no control. This velocity is determined by bubble size and water temperature, in other words, by the features of the equipment installed. The total volume and size of air bubbles produced during depressurization are proportional to the stream's pressure, the flow rate and the pressure-reducing device (nozzle). Large air bubbles produce a rapid and turbulent rise velocity, which reduces the air-solid contact time, as well as the bubble's specific surface. More efficient solid removal can be achieved with smaller air bubbles, which increases the contact time and the total surface developed by the bubbles.

6.5.3. *Contact time*

Increasing the contact time improves the performance. The design should be based on the contact time in relation to the nominal flow so that the performances delivered are in line with the ones expected at all times.

6.5.4. *Bubble volume concentration*

The most important operating and control variable affecting DAF performance is bubble volume concentration in the contact zone. Air concentration can be modified by changing the saturator pressure or by modifying the recirculation rate. However, since the saturator pressure does not vary greatly, only the bubble volume concentration can be modified by intervening in the recirculation flow rate. As indicated in Table 6.3, a variation in this rate affects bubble volume on the condition that the equipment is provided for this purpose.

An even distribution of saturated water across the flotation tank's width is equally important. An adjustment of the recycle flow rate matching the raw water flow and quality is also recommended. A bubble diameter between 20–100 μm is recommended in relation to the saturation pressure.

Air bubbles		Flocs	
Diameter (μm)	20–100	Diameter (μm)	100–300
Mean diameter (μm)	50	Mean diameter (μm)	100
Concentration (bubbles·m^{-3})	4×10^9–2.7×10^{11}	Floc density (kg·m^{-3})	1,005–1,010
Air dosage (L·m^{-3})	5–10	Floc concentration (flocs·m^{-3})	2.5×10^7–1.9×10^8

Table 6.6. *Characteristics of air bubbles and flocs*

The nozzles connected to the floor can work with a fixed flow or with a manually adjustable flow. The advantage of the latter is that the operator can modify the flow rate, if needed.

The formation of floc–bubble couplings can be controlled depending on their respective characteristics.

6.5.5. Gas solubility

Gas solubility ensures a sufficient air flow under optimal P and T°C conditions. After the starting up of saturators, the time required to reach a steady-state composition of air can take several hours. For 5 bar saturator pressures and more and saturation efficiencies of at least 85%, the steady-state composition of air is reached in less than 4 h. For large-scale installations with continuous saturator operation, it is the steady-state composition of air (the composition of air inside the saturator at equilibrium with the saturator's outlet water) that is interesting.

Power consumption is 3–5 kWh/10,000 m^3·d, higher than that of settlers, which is approximately 0.5–1.2 kWh/10,000 m^3·d.

6.5.6. Hydraulic efficiency

The average water losses caused by sewage sludge discharge may vary between 1.3 and 2%.

6.5.7. Air/water ratio

The amount of air per cubic meter of treated water results from:

$$\frac{A}{V} = \left(\frac{((e(P + 101.3) - Patm)}{H} \right) \cdot \frac{R}{(1 + R)}$$

where:

– A/V: air/water ratio (g of air·m^{-3} of water);

– e: saturator efficiency;

– P: relative pressure (kPa);

– $Patm$: atmospheric pressure (kPa);

– H: Henry's constant (kPa·mg^{-1}·L);

– R: recirculation rate.

6.6. Performance and monitoring

DAF is a robust solution for the treatment of water from dams, ponds, reservoirs, seawater, water rich in algae and/or highly colored water, provided that the design and operating conditions are properly respected. This process easily accepts stop-start conditions, increased flow rates and certain variations in water quality. One of its main advantages is that it can start within 30 min and can accept being out of service for several hours, without any deterioration in water quality.

In ozone injection flotation systems (see Veolia technologies: Ozoflot® and Flottazone®), the oxidation provided by ozone is useful for removing algal toxins, substances responsible for tastes and odors, etc.

The additional benefit is that ozone helps immobilize motile algae species and dissociate heavier colonies, which facilitates their removal by flotation.

6.6.1. *Treatment monitoring*

In the field, the following measurements provide information on the quality of the treatment performed:

– continuous turbidimeter on floated water;

– surface condition of the whitewater injection zone and of the sludge blanket;

– floc formation in the flocculation tank;

– consumption of reagents, air and energy.

At the laboratory, treatment monitoring concerns one or more of the following parameters (depending on the application):

– raw water and floated water turbidity;

– raw water temperature;

– suspended solids and organic matter in raw water and floated water;

– dry matter in sewage sludge (beware of dissolved matter, especially in seawater) and sludge flow;

– raw water, coagulated water and floated water pH;

– raw water and floated water color;

– dosage of coagulation reagents;

– verification of residual coagulant (iron or aluminum) in floated water;

– algae count;

– silt density index (SDI).

6.6.1.1. *Residual turbidity (NTU)*

A well-run flotation unit typically produces a floated water quality between 0.9 and 2.0 NTU, with some exceptional examples averaging 0.5 NTU.

The use of mechanical scrapers or the start-up of a plant can cause short-term disturbances of up to 2.5 NTU, but this is usually short lived and the plant quickly returns to good quality conditions.

The residual turbidity at the outlet of a DAF varies depending on the whitewater injection pressure and on floc diameter (µm). Figure 6.28 clearly shows that, for a 5 bar pressure, residual turbidity is less than 2 NTU for a 50 µm floc diameter. On the other hand, for a 3 bar pressure, residual turbidity is higher for the same floc diameter.

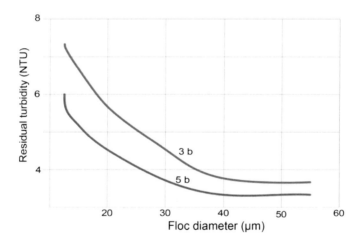

Figure 6.28. *Residual turbidity as a function of floc diameter (raw water turbidity: 15–20 NTU). For a color version of this figure, see www.iste.co.uk/gaid/watertreatment1.zip*

Flocs smaller than 10 µm hardly adhere to bubbles, and their removal is very low, as the particle size approaches 1 µm. It is therefore essential for chemistry, coagulation and flocculation to be effective in converting (notably colloidal) particles to a size far over 20 µm.

A hydraulic short circuit in the separation zone results in a diversion of part of the flow toward the bottom of the flotation unit and naturally toward the floated water outlet. This provokes a deterioration in water quality and an immediate increase in turbidity.

Turbidity removal varies with temperature, and at low temperatures (<5 °C), water density increases while degrading separation efficiency. Coagulation efficiency (based on a residual coagulant) may also diminish due to a change in optimal pH conditions as well as a decrease in the reactivity of certain metallic coagulants. In these circumstances, minor pH corrections are required to restore optimum conditions. In particular, the addition of flocculants (polymers) helps maintain water quality and flow.

6.6.2. Performance of DAF systems in relation to algae removal

Flotation is generally considered the best technology available for algae removal. At DAF, micro air bubbles adhere to algae particles and drag them to the surface, where they are collected and evacuated. However, the addition of a coagulant is necessary for the algae–floc–bubble aggregate to form. This coagulant can be an aluminum or an iron salt, whose dosage depends on the concentration and type of algae present in the resource. Even if the dosages are lower than the ones admitted in a settler, their injection is essential and can be accompanied by the injection of polymer when algae concentrations are too important to handle.

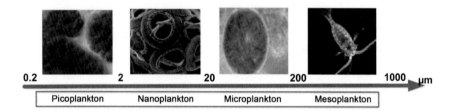

Figure 6.29. *Types of plankton easily removed by flotation. For a color version of this figure, see www.iste.co.uk/gaid/watertreatment1.zip*

An air/SS ratio of 0.03 (g·g^{-1}) and at least 8 mg of air per liter of water are needed. A DAF operating with a hydraulic loading rate of 30 m·h^{-1}, or 720 m·d^{-1}, and an algae concentration of 50 million units·L^{-1} leads to an algae applied load of 3.6×10^{13} units·d^{-1}. For 99.6–99.8% efficiency, the removal of motile algae species can be improved by their inactivation with chemical preoxidation with potassium permanganate, chlorine dioxide or ozone. The formation of a hydrophobic surface facilitates their separation. However, this preoxidation often induces an increase in dissolved organic carbon, which must also be removed at the DAF level.

It is advisable to responsibly control residual Mn (with a $KMnO_4$ preoxidation), as well as toxin release with O_3 and ClO_2, due to the cell lysis provoked by the addition of these two oxidants.

6.6.3. Performance against parasites

General rules require a cumulative credit of 2.5–3 log for the removal of *Cryptosporidium* oocysts by flotation and filtration. This is equivalent to the value required for settling and filtration. Although the exclusive application of DAF has resulted in the removal of up to 2.5 log oocysts, it can reach 4 log when combined with filtration.

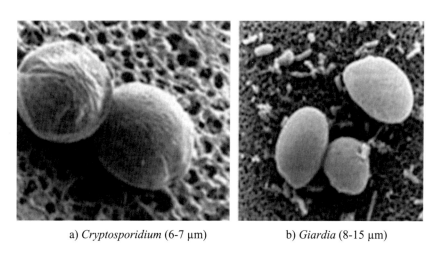

a) *Cryptosporidium* (6-7 µm)   b) *Giardia* (8-15 µm)

Figure 6.30. *Cryptosporidium (a) and Giardia (b) parasites*

6.6.4. Performance with the addition of PAC

It is possible to add PAC to DAF units. This is done with a view to eliminating (humic and fulvic) organic substances, the molecules responsible for tastes and odors (methyl isoborneol and geosmin), color, algal toxins, pesticides and other micropollutants. Dosages are variable between 10 and 60 mg·L^{-1}, depending on the nature and concentration of the substances to be removed. Dosages also depend on operating conditions such as contact time, nature of the PAC and pH. In an acid medium, which has a positive effect on PAC adsorption mechanisms, dosages are lower than in a medium with basic pH.

Figure 6.31 shows pictures taken with an electron microscope of a PAC surface before and after its injection into a DAF unit.

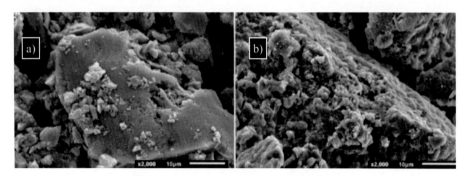

Figure 6.31. *Addition of PAC (Cabot-Norit) in a DAF unit:*
(a) before introduction; (b) after introduction

6.7. Veolia technologies using flotation

The Veolia technologies using flotation applicable to drinking water are as follows:

– Spidflow®;

– Spidflow® filter;

– Ozoflot®;

– Flottazone®.

6.7.1. *Spidflow®: principle*

The Spidflow® procedure constitutes a DAF process. In particular, it involves an indirect flotation process, whereby a part of the clarified water is recirculated under pressure and brought into contact with the air to produce microbubbles. These microbubbles give the mixture a milky appearance, which explains the name "whitewater" when referring to this type of recirculation. Flocs are brought to the structure's surface thanks to these microbubbles.

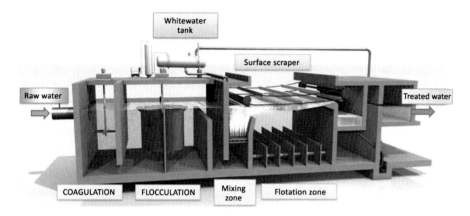

Figure 6.32. *Diagram of Spidflow®'s principle. For a color version of this figure, see www.iste.co.uk/gaid/watertreatment1.zip*

The previously coagulated and flocculated raw water enters through the lower portion of the contact zone (also called the "mixing zone"). There, the flow is homogenized thanks to a perforated floor. Pressurized water is introduced above this floor with the help of uniformly distributed nozzles specifically designed for this purpose. Whitewater is created by release. Flocs and microbubbles agglomerate in this mixing zone. The flow then enters the separation zone (also called the "flotation zone"), and these agglomerates rise to the structure's surface. Figure 6.33 illustrates a view of aggregates. Part of the flow is recirculated for the production of whitewater.

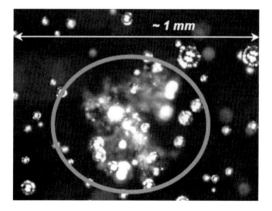

Figure 6.33. *Floc–bubble aggregates*

The passage of the water to be treated through the bubble clouds makes it possible to create floc–bubble contacts, which will in turn form aggregates that rise to the surface. Through the density of the nozzles installed on the floor, the homogeneous distribution of water neutralizes the hydraulic short circuits, which can lead to a loss of efficiency due to a lack of contact between the particles in suspension and the air bubbles.

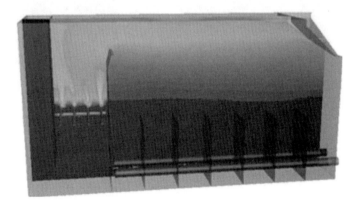

Figure 6.34. *Spidflow® fluent modeling operating at 35 m·h⁻¹. For a color version of this figure, see www.iste.co.uk/gaid/watertreatment1.zip*

The nozzles are arranged vertically so that pressurized water is pulverized following a co-current direction with the water to be treated in the upward current. The production of whitewater is performed in two alternative ways:

– a first solution comprises a recirculation pump, a prefilter and a saturator;

– A second solution comprises a prefilter and a pump producing whitewater directly. This solution is particularly applied for small installations.

The pressurization system used makes it possible to reach the water saturation point at pressures greater than 4 bar at the corresponding temperature and salinity. Moving upward toward the surface, the floc–bubble aggregates can shear as they reach the interface. This can be avoided by widening the contact zone's vertical wall by an angle of a few degrees. The mirror velocity is greatly reduced, as is the turbulence, mitigating the possible breakage of aggregates.

Spidflow® includes a mobile scraper that moves across the upper portion of the separation zone, removing not only air bubble beds but also the accumulated sludge.

The mechanical scraper works in a rhythmic way and evacuates the sludge toward a trough arranged for this purpose. The scraper is designed to limit the height and the number of skimmer blades per scraper to respect a velocity of less than 180 linear meters per hour.

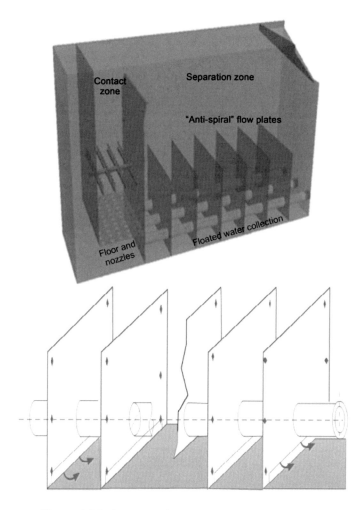

Figure 6.35. *Anti-spiral flow plates. For a color version of this figure, see www.iste.co.uk/gaid/watertreatment1.zip*

The separation zone has vertical anti-spiral flow plates installed at the recovery zone (or floated water collection zone). These plates prevent the formation of recirculation loops in the flotation zone and virtualize the flow to ensure proper separation between the rising agglomerates and the descending clarified water. These plates, as well as the collection lines, ensure the stability of the microbubble bed to consolidate the bubble–floc agglomerate and to facilitate clarification. They prevent short circuits and avoid the escape of SS into the treated water.

Please note that in the presence of plates (Figure 6.36), the flow tends to stratify naturally in the separation zone. The plates measure between 30 and 300 cm high and are spaced at a 20–300 cm distance. These plates are essential for structures longer than 5 m.

The clarified water is uniformly collected at the bottom of the structure due to perforated pipes, which all reunite in a common collector.

If the structures are placed in the open, the flotation zone is covered to prevent bad weather from breaking the sludge layer.

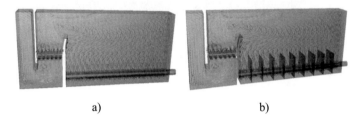

a)   b)

Figure 6.36. *Fluent modeling without plates (a) and with plates (b) at 30 m·h^{-1}. For a color version of this figure, see www.iste.co.uk/gaid/watertreatment1.zip*

6.7.1.1. Spidflow® application domains

For drinking water purposes, Spidflow® is used for the clarification of surface water (rivers, lakes, reservoirs). For desalination, Spidflow® is properly suited as an advanced pretreatment stage before granular or membrane filtration. Spidflow® is highly suitable for the production of process water. Finally, Spidflow® is also effective for waters with the following characteristics:

– slightly turbid water: the design is based on maximum average values of 50 NTU. These recommendations need to be adjusted depending on the nature of SS; for example, colloidal particles can cause high turbidity without producing much SS;

– colored water rich in humic substances;

– water rich in algae, and in organic matter in general;

– water containing low-density particles that are difficult to agglomerate;

– sea water.

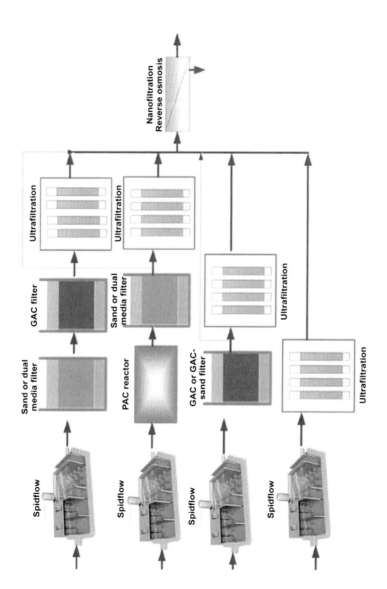

Figure 6.37. *Spidflow® position at a water treatment plant. For a color version of this figure, see www.iste.co.uk/gaid/watertreatment1.zip*

6.7.1.2. *Integration into the water treatment plant*

Spidflow® can be integrated into a water treatment plant, depending on the quality of the raw water.

It can be set up:

– either upstream of a mono media or dual media filter (sand, GAC, anthracite-sand, pumice-sand, GAC-sand, sand-MnO_2, GAC-MnO_2);

– or upstream of a three-medium filter (anthracite or pumice-sand-MnO_2 or GAC-sand-MnO_2);

– upstream of an activated carbon reactor (powdered or micrograin);

– upstream of a microfiltration or ultrafiltration membrane.

6.7.1.3. *Design parameters*

6.7.1.3.1. Retention time

Typical retention times, calculated according to the nominal flow rate, are shown in Table 6.7.

	Temperatures ≤5°C	5° < Temperatures ≤10°C	Temperatures >10°C
Coagulation time	4 min (by mechanical impeller)	2 min (by mechanical impeller)	1 min (by mechanical impeller or static mixer)
Flocculation time	Without Turbomix: 30 min With Turbomix: 20 min	Without Turbomix: 20 min With Turbomix: 10 min	Without Turbomix: 10 min With Turbomix: 5 min
Separation time in the flotation zone	>8 min	>6 min	

Table 6.7. *Retention time throughout the different stages of Spidflow®*

6.7.1.3.2. Mirror velocity

Spidflow® design is calculated at the maximum flow rate to guarantee a hydraulic loading rate of up to 40 m·h^{-1}. In some cases of traditional surface water, the velocity may be higher.

Temperatures	Types of water	Maximum mirror velocities
Temperature < 10°C	Lightly loaded and loaded water	30 m·h^{-1}
Temperature > 10°C	Heavily loaded water (turbidity > 50 NTU or DOC > 8 mg·L^{-1})	30 m·h^{-1}
	Lightly loaded water (turbidity < 10 NTU)	40 m·h^{-1}

Table 6.8. *Design velocity implementation conditions*

The recirculation rate represents 5–10% of the operating flow. The range of structures covers unit flows of up to 3,640 m^3·h^{-1} (for 10 m wide × 10 m length structures, at 40 m·h^{-1} including the recirculation flow).

6.7.1.4. *Operating parameters: consumption and reagent dosage*

6.7.1.4.1. Coagulants

Types of water	Typical iron or aluminum values
Surface water (NTU turbidity < 5 NTU)	0.5–1.5 ppm
Contaminated surface water (10–60 ppm of SS)	3–6 ppm
High load water (SS or algal bloom)	6–9 ppm
Surface water with dissolved organic carbon	2–30 ppm

Table 6.9. *Coagulant dosage (estimated in metal mg·L^{-1})*

6.7.1.4.2. Polymer

	Temperatures < 10°C	Temperatures > 10°C
Anionic polymer	0.1–0.4 mg·L^{-1}	Only during algal blooms or highly turbid surface water (0.2–0.5 mg·L^{-1})

Table 6.10. *Flocculant dosages*

An in-line dosage of PAC can be injected upstream of the coagulation tank for the removal of algal toxins, organic matter or micropollutants. A dosage of 10–20 mg·L^{-1} is generally expected, but it depends on the type of contaminant and its concentration. Spidflow®'s power consumption is generally between 30 and 45 Wh·m^{-3} of treated water for a 10% recirculation. The recirculation pump represents the largest energy-consuming device (70–90%).

6.7.1.5. Spidflow® performances

	Units	Typical values
Turbidity	NTU	0 to 50 and up to 100 NTU episodically
SS	mg·L^{-1}	0 to 70 and up to 200 NTU episodically
Color	°H or PtCo	0 to 300
SDI3		No limits. Spidflow® is relevant for SDI3 > 30
TOC	mg·L^{-1}	0 to 20
Algae	Cell·L^{-1}	No limits Typical values:100,000–2,000,000 cells per L during blooms
Chlorophyll		No limits Presence of algae with 8–10 µg·L^{-1} chlorophyll levels Peaks can reach 20–30 µg·L^{-1}, or even 60 µg·L^{-1} at the most
Oils and hydrocarbons	mg·L^{-1}	Effective for free and emulsified oils. Reduced efficiency for dissolved oils or hydrocarbons

Table 6.11. *Typical raw water quality*

	Units	Typical values
Turbidity	NTU	<2 NTU if raw water turbidity <20 NTU <3 NTU if raw water turbidity <100 NTU
SS	mg·L^{-1}	<5 mg·L^{-1} if SS in raw water <50 mg·L^{-1}
Apparent color	°H or PtCo	<15
TOC	mg·L^{-1}	Reduction of approximately 15–50%*
Algae		Reduction of 90–99.9%
Oils and hydrocarbons	mg·L^{-1}	Reduction >90% if dissolved, >95% for free or emulsified oils or hydrocarbons

*Depends on coagulation and flocculation conditions (reagent dosage, flocculation pH).

Table 6.12. *Typical floated water quality*

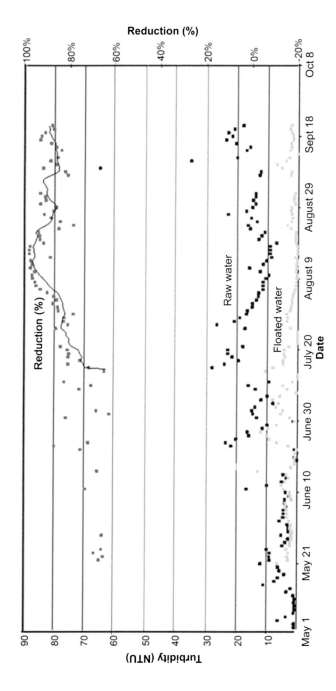

Figure 6.38. *Turbidity reduction using Spidflow®. For a color version of this figure, see www.iste.co.uk/gaid/watertreatment1.zip*

Figure 6.38 shows the evolution of turbidity reduction with the use of Spidflow®
from the start-up phase to the equilibrium phase. A reduction higher than 80–85% is
obtained with a residual turbidity always lower than 2 NTU, even for an inlet
turbidity oscillating between 10 and 25 NTU.

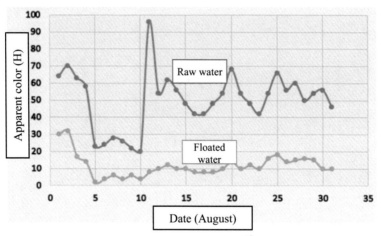

Figure 6.39. *Color removal using Spidflow®. For a color version
of this figure, see www.iste.co.uk/gaid/watertreatment1.zip*

Figure 6.39 shows the removal of apparent color using Spidflow® for a month. A
reduction higher than 80–90% is obtained with a residual variable color between 5
and 18 and a value <15 °H, more than 90% of the time.

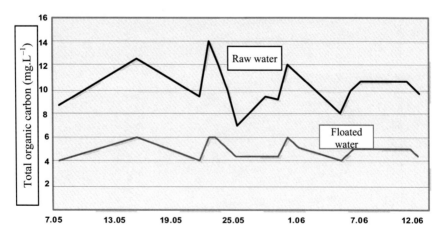

Figure 6.40. *TOC removal using Spidflow®. For a color version
of this figure, see www.iste.co.uk/gaid/watertreatment1.zip*

The removal of organic matter primarily depends on the dosage of coagulation reagents and on coagulated water pH. The ferric chloride ($FeCl_3$, 41%) dosage was 160 mg·L^{-1}, and the average coagulation pH was 6.5. The reduction results presented in Figure 6.40 show an average reduction of 55% after using Spidflow®.

Units		Typical values
Floating sludge concentration	g·L^{-1}	20–35
Water losses in the flotation unit	%	0.1–0.5

Table 6.13. *Other indicator parameters*

6.7.1.6. Spidflow® performance in relation to algae removal

Algal blooms are generally associated with high concentrations of nutrients (phosphorus and nitrogen). Their growth mainly spans the surface water layers until reaching a depth where there is sufficient light. Other parameters responsible for algae development are high temperature (above 15°C) and low water circulation (long retention time, favorable weather conditions), as observed in ponds, reservoirs and dams.

6.7.1.6.1. First case: water retainer

On average, cell and algae reductions have a 99.8% success rate, that is, 2–4 log. The number of cells in treated water was between 0.5 and 90 cells per m·L, mostly *Radiocystis* from the Cyanobacteria family. Reduction seems to be poorly selective and is similar for most families.

The maximum amount of algae observed was 13,760 algae·mL^{-1}, in other words, 104,134 cells. Populations are varied, comprising species such as Cryptophyceae, Chrysophyceae, Chlorophyceae and Cyanobacteria, which predominate alternately. The bloom is mainly attributed to Chlorophyceae. Algae reductions are astonishingly high: 99.95% on average, barely lower than 4 log. The number of cells in treated water was lower than 50 cells·mL^{-1}.

6.7.1.6.2. Addition of PAC with Spidflow®

In the presence of organic matter, micropollutants or algal toxins (either periodically or frequently), PAC can be added upstream of Spidflow®. When these substances are present periodically or seasonally, their concentration is not significantly high (TOC < 3 mg·L^{-1} and total micropollutants < 0.8 µg·L^{-1}), activated carbon is added upstream of the flotation unit, with variable doses between 10 and 25 mg·L^{-1}.

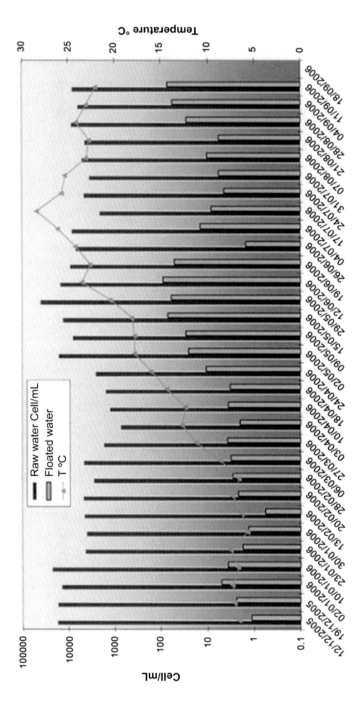

Figure 6.41. *Algae removal performances using Spiflow second case: pond. For a color version of this figure, see www.iste.co.uk/gaid/watertreatment1.zip*

Figure 6.42. *Algae removal using Spidflow® on a pond (Ouest Bretagne). For a color version of this figure, see www.iste.co.uk/gaid/watertreatment1.zip*

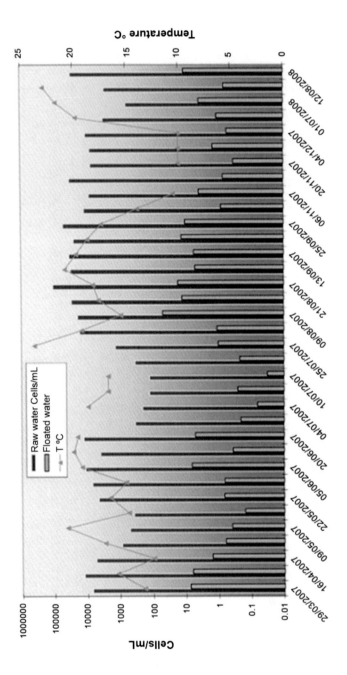

Figure 6.43. *Algae removal performances using Spidflow®. For a color version of this figure, see www.iste.co.uk/gaid/watertreatment1.zip*

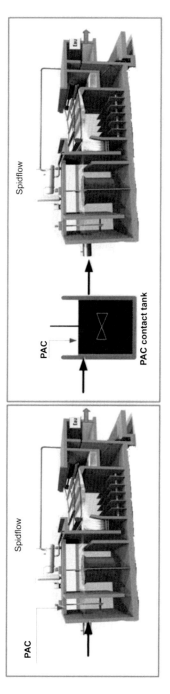

Figure 6.44. *Activated carbon injected into the coagulation tank or with a contact tank upstream of Spidflow®. For a color version of this figure, see www.iste.co.uk/gaid/watertreatment1.zip*

When these substances are frequently present and their concentration is much higher (TOC between 3 and 4 mg·L^{-1} and higher than 2 μg·L^{-1} for micropollutants), PAC is introduced into a contact tank for 10–15 min (Figure 6.44). The type of PAC is chosen depending on the substances to be removed, and their size can be between 6 and 35 μm.

Flota test trials (Figure 6.45) are recommended to determine the type and concentration of PAC to be injected.

Figure 6.45. *Flota-test trials with PAC: (a) flocculation and (b) flotation. For a color version of this figure, see www.iste.co.uk/gaid/watertreatment1.zip*

The application of PAC on plants equipped with Spidflow® is effective for the reduction of organic substances and micropollutants. Figure 6.46 shows the flotation zone where injected activated carbon has accumulated.

Figure 6.46. *Powdered activated carbon (Jacobi) recovered from Spidflow®'s surface. For a color version of this figure, see www.iste.co.uk/gaid/watertreatment1.zip*

Turbidity at the outlet is practically identical to that observed without the addition of PAC. The rate of pesticides (atrazine, desethylatrazine [DEA] and desisopropylatrazine [DIA]) at the outlet was < 0.1 $\mu g \cdot L^{-1}$ for a PAC (Jacobi) dose of 10 $mg \cdot L^{-1}$. The microcystin concentration was <1 $\mu g \cdot L^{-1}$.

6.7.2. *Advantages and limitations of DAF systems*

Table 6.14 summarizes the advantages and limitations of the Spidflow® flotation unit.

Advantages	Limitations
Compact process due to high mirror velocities (25–40 $m \cdot h^{-1}$) Algae clarification efficiency and hydrocarbon removal	Need to cover DAF systems to avoid sludge breakage in case of rain
Highly restrained (practically zero) use of polymer	Higher energy consumption than settlers (approximately 30–50 $Wh \cdot m^{-3}$), but still lower for large installations
The production of highly concentrated sludge (2–3% on average, if sludge is evacuated by surface scraping), which does not require thickening before dehydration, therefore inducing low water losses	Limited inlet turbidity (<10 NTU, even if 50 NTU peaks can be accepted)
Operating flexibility and compactness	Saturator and nozzle floor are necessary
The possibility of using powdered activated carbon and removing it by flotation to improve the reduction of organic matter, endocrine-disrupting compounds or pesticides	

Advantages (cont'd)	Limitations (cont'd)
Extracted sludge concentration 2–3%	
No polymers, only in case of algal blooms or high turbidity	
Highly restrained coagulant dosage, except in case of organic matter removal	
Flocculation contact time lower than for settling	
Very good algae removal (>99.5%)	
Very good parasite removal (Cryptosporidium and Giardia) 2–2.5 log	

Table 6.14. *Spidflow® advantages and limitations*

Figure 6.47. *Photos of treatment plants equipped with Spidflow®: (a) Eemshaven (the Netherlands), (b) Marmara (Turkey), (c) Narva (Estonia), (d) Kermorvan (France), (e) Trégat (France), (f) Toulon la Valette (France). For a color version of this figure, see www.iste.co.uk/gaid/watertreatment1.zip*

6.7.3. *Spidflow® filter*

Spidflow® filter is a combination of several physical and chemical single stages to clarify water. It comprises a coagulation and a flocculation stage, as well as a flotation stage followed by a filtration stage.

The coagulation-flocculation stages and flotation are similar to those described for the Spidflow® process. Particles are destabilized after the injection of a coagulant, and small flocs are consolidated by flocculation. The floc–air bubble aggregates are then formed in the contact zone after the injection of pressurized water (4–8 bar) through nozzles. Floc–bubble collisions create aggregates, which are then brought to the surface due to the rise velocity. Optimal flotation performances are obtained at a pressurized water flow rate of 5–10% of the feed rate. These mechanisms are described in detail in the previous chapters. The sludge accumulated on the surface is periodically removed by scraping. Floated water feeds the dual media filter by gravity. The filtering materials are anthracite or pumice stone for the upper-level media and sand for the lower layer media.

The operation of this filter is equivalent to the dual media filters, as described in Chapter 8, Volume 2. The filter was regularly backwashed with air and filtered water. Different types of water can be used for backwashing (filtered water or brine in the case of desalination plants comprising reverse osmosis membranes).

When backwashing only with water, an additional floc removal stage is added to accelerate the removal of flocs that are present in the flotation zone.

Due to the medium's significant height (3 m in total), biological activity thrives within the filter, often degrading biodegradable organic molecules and clogging up membranes. In addition, this filtration height induces long filtration cycles due to the good and constant quality of floated water, a feature that also promotes biological activity.

Alternatively, timers can be used (open/closed duration valve). The washing sequence is managed taking three parameters into account: production time, production volume and increase in head loss.

This sequence includes an operating mode similar to the one described for filters (see Volume 2, Chapter 8). This entails an adjustment of the water level up to 5–10 cm above the media.

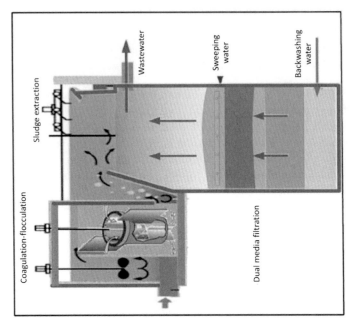

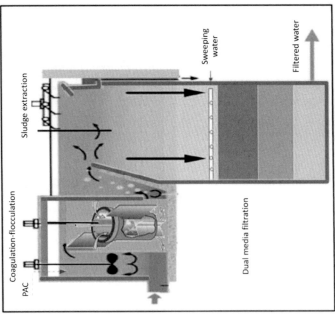

Figure 6.48. *Spidflow® filter functioning and backwashing. For a color version of this figure, see www.iste.co.uk/gaid/watertreatment1.zip*

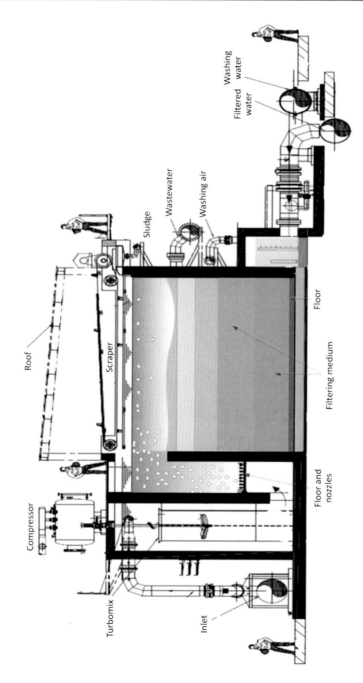

Figure 6.49A. *Diagram of Spidflow® filter. For a color version of this figure, see www.iste.co.uk/gaid/watertreatment1.zip*

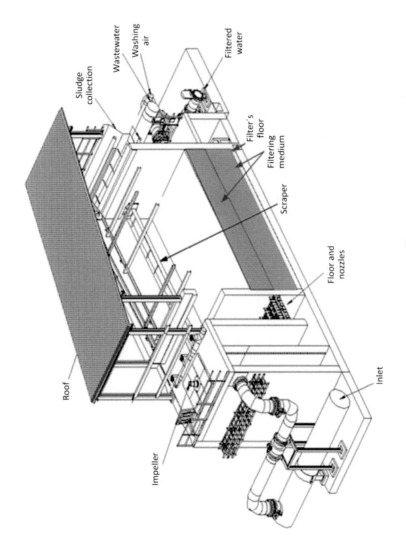

Figure 6.49B. *Diagram of Spidflow® Filter. For a color version of this figure, see www.iste.co.uk/gaid/watertreatment1.zip*

Then, it follows a washing sequence only with air, followed by the air + water phase, without draining wastewater.

Finally, a washing sequence was performed with water alone, at a high velocity and with a 20–25% expansion of the light media (upper layer media).

Mini-washes are periodically carried out during production to prevent the risk of embolism within the media's heart due to the trapping of residual air bubbles. This mini-phase lasts between 10–20 s.

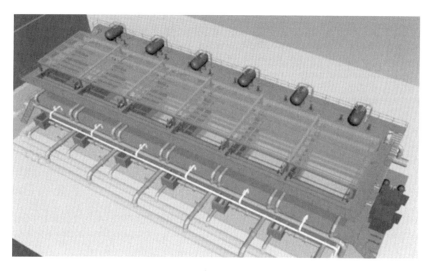

Figure 6.50. *View of six Spidflow® filters working in parallel. For a color version of this figure, see www.iste.co.uk/gaid/watertreatment1.zip*

6.7.3.1. *Spidflow® filter position at a water treatment plant and Spidflow® filter performances*

Spidflow® filter performances show that the turbidity of filtered water is well below 0.5 NTU, at values approaching 0.2 NTU.

Coagulant dosage (with pH control) eliminates up to 50% of the total organic carbon. In addition, PAC can be injected upstream or at the structure's inlet to remove organic matter, color, pesticides and other micropollutants.

Algae removal is identical to that obtained with Spidflow® on its own because the reduction mechanisms are equivalent.

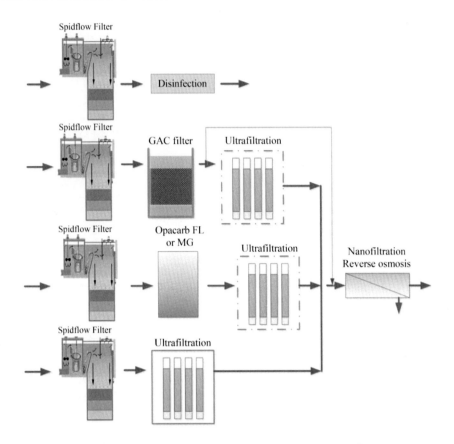

Figure 6.51. *Various possibilities for integrating the Spidflow® filter into a water treatment plant. For a color version of this figure, see www.iste.co.uk/gaid/watertreatment1.zip*

	Units	**Values**
Turbidity	NTU	<0.5 NTU
SS	–	<1 mg·L^{-1}
True color	°H or PtCo	<5
TOC	mg·L^{-1}	15–50% removal*
Algae	–	Removal > 99.8%
Oils and hydrocarbons	mg·L^{-1}	Removal > 90% for dissolved substances > 95% for free or emulsified substances

*Depends on the coagulation-flocculation operating conditions.

Table 6.15. *Quality of outlet filtered water using a Spidflow® filter*

6.7.3.2. *The advantages of the Spidflow® filter*

Spidflow® has many advantages, such as:

– a compact design with a limited footprint (gain of at least 25% compared to the conventional design);

– no thickening needed: highly concentrated floating sludge (>2%), avoiding the additional thickening stage commonly used in sludge treatment;

– modularity: can be installed in treatment plants of all sizes;

– rapid reaction to variations in raw water quality, reinforced by optimized hydraulic conditions and Veolia's exclusive whitewater injection nozzles;

– reduced backwashing frequency thanks to an optimized management of filtration operations;

– the guarantee of production at the treatment plant's full-time operating capacity;

– the removal of clogging substances: the Spidflow® filter is the ideal pretreatment solution for removing low-density suspended solids, algae, oils and hydrocarbons, as well as soluble organic compounds that are detrimental to the membranes arranged downstream;

– lower capital and operating costs;

– compact design, which reduces equipment and construction costs;

– energy efficient design;

– optimal consumption of chemical reagents.

6.7.4. *Ozoflot®*

The effectiveness of mass transfer from the gas phase to the liquid phase depends on the characteristics of the transfer equipment, the kinetics of ozone degradation in water and the number and size of bubbles produced. The effectiveness of ozone as an oxidant or as a disinfectant can be increased by creating a larger surface/volume ratio through the generation of small bubbles.

Several mechanisms are engaged during an ozonation reaction in water containing organic matter. These involve a direct chemical reaction of ozone with organic matter, in the role of an electron acceptor, as well as an indirect reaction, through the generation of a strongly oxidizing hydroxyl radical.

Once ozone is dissolved into water, it quickly decomposes through a series of reactions:

$$OH^- + O_3 \rightarrow H_2O + O_2^-$$

$$2H_2O \rightarrow 4H^+ + O_2^-$$

$$O_2^- + O_3 \rightarrow O_2 + O_3^-$$

$$O_3^- + H^+ \rightarrow HO_3$$

$$HO_3 \rightarrow O_2 + OH^\circ$$

$$O_3 + OH^\circ \rightarrow HO_4$$

$$HO_4 + HO_3 \rightarrow H_2O_2 + O_2 + O_3$$

However, the chemical composition of water can influence the kinetics of these reactions. This happens because the reaction chain is initiated by the hydroxyl ions present in the water with the generation of intermediate compounds, such as HO_2, HO_3, and HO_4, which can trigger other reactions. The decomposition of ozone into secondary oxidants, such as the hydroxyl radical (OH°), takes place very quickly at high pH values. While oxidation by ozone is selective, oxidation by the OH° radical (which is a non-selective oxidant) is more effective than ozone.

The Ozoflot® process combines oxidation reactions with ozone and promotes floc flotation via the gas bubbles released, which become attached to them. The attraction between gas bubbles and flocs is the result of various physical and physicochemical forces, as described in the previous paragraphs.

The flotation of dispersed air involves the formation of bubbles by diffusers. In most cases, large bubbles are formed (sizes 300–900 μm), which can be effective for the treatment of suspensions containing large particles, such as the separation of minerals and the processing of industrial waste. However, dispersed air flotation has generally proven unsuitable for water treatments involving the removal of fine particles.

The Ozoflot® process benefits from specific diffusers, and ozone-treated air is injected through porous plates swept by a water stream to generate fine bubbles and improve the efficiency of ozone transfer. The bubble dimensions are between 100 and 500 μm.

Before entering the ozone flotation unit, water undergoes coagulation and a short flocculation. Floating sludge is evacuated by raising the water level.

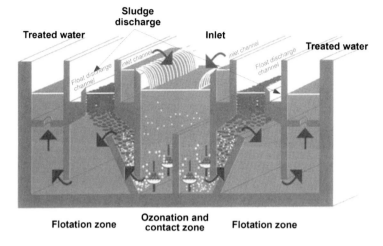

Figure 6.52. *Ozoflot® process. For a color version of this figure, see www.iste.co.uk/gaid/watertreatment1.zip*

Figure 6.53. *Transfer of O₃ through specific porous supports. For a color version of this figure, see www.iste.co.uk/gaid/watertreatment1.zip*

Table 6.16 summarizes the design parameters.

We thus obtain:

– water particle removal by means of flotation via the formation of flocs resulting from the addition of a coagulant;

– algae removal (>99%);

– oxidation of organic substances (30–50%, depending on operating conditions);

– color removal (30–50%, depending on operating conditions);

– reduction of microorganisms.

Parameters	Values
Mirror velocity (m·h^{-1})	25
Velocity in the O$_3$ contact zone (m·h^{-1})	13
Acceptable turbidity (NTU)	10
O$_3$ dosage (mg·L^{-1})	2–4
Sweep water flow	7–10% of inflow

Table 6.16. *Ozoflot$^®$ process design parameters*

6.7.5. *Flottazone$^®$*

The purpose of the Flottazone$^®$ process is to combine the oxidation properties of ozone and physical flotation within the same tank. Physical flotation demands "calibrating" bubbles, as well as their number. The bubbles generated depend on gas solubility and dissolution pressure (the size of the microbubbles produced decreases with dissolution pressure). Ozone dissolution into process water depends not only on the concentration of ozone in the gas to be injected but also on the process water flow rate.

The Flottazone$^®$ process generates fine ozone bubbles by pressurizing water and then injecting ozone into it. We thus obtain ozone-treated water by pressure, which it is possible to relieve at the level of a flotation device, thereby producing numerous ozone fine bubbles. This process resembles air flotation but is specific due to the use of ozone.

The addition of the coagulant upstream of the flotation column makes it possible to destabilize, separate and accumulate particles at the level of the flotation unit's surface, as occurs with traditional flotation. However, the addition of ozone improves flocculation and favors better separation by flotation.

In the Flottazone$^®$ process, ozone dissolution is performed in-line using a hydroinjector followed by a static mixer. The Venturi makes it possible to suck ozone-treated gas and to inject it into pressurized process water. At the same time, this system carries out a first ozone dissolution stage due to the intimate mixture created at the suction level.

The static mixer located downstream of the Venturi completes the gas/liquid mixture and ozone transfer. The ozone-treated gas + process water mixture enters a reactor working under pressure to separate the ozone-treated process water from the gas poor in ozone. The outlet of the gas/liquid separation reactor is connected to a set of ozone-treated process water releasing nozzles, and it is at this level that fine ozone bubbles are generated.

The fine bubble generation nozzles are placed at the flotation zone's bottom. The coagulant is injected upstream into the raw water. Flocculation has a contact time of 3 min. In its lower portion, the flotation tank is equipped with troughs for collecting the treated water and for discharging any sludge that could have settled at the structure's bottom. In the upper portion, troughs enable the overflow evacuation of any floating materials that could have remained at the water level.

We thus obtain:

– water particle removal by means of flotation via the formation of flocs resulting from the addition of a coagulant;

– algae removal (>99%).

Oxidation of organic substances (30–40%, depending on operating conditions):

– color removal (30–40%, depending on the operating conditions);

– reduction of microorganisms.

Parameters	Values
Coagulation	In-line
Flocculation (min)	3
Mirror velocity ($m \cdot h^{-1}$)	15–30
Driving pressure (bar)	7
Gas/liquid ratio	10–30%
O_3 concentration in gas ($g \cdot m^{-3}$)	80–150
Transfer efficiency	90%
O_3 concentration in water ($mg \cdot L^{-1}$)	1.6–2
Acceptable turbidity (NTU)	10

Table 6.17. *Flottazone® process design parameters*

O_3 transfer	Contact time (s)	O_3 concentration/gas $(g \cdot Nm^{-3})$	Residual O_3 $(mg \cdot L^{-1})$		O_3 rate $(mg \cdot L^{-1})$
			Inlet	Outlet	
Injector	0.1	100	0	10.2	19.2
Static mixer	0.3	–	10.5	13.8	–
Separator reactor	45	–	13.7	14.6	–
Flottazone®	900	–	1.8	0.4	1.8

Table 6.18. O_3 operating conditions
at an active water treatment plant

6.7.6. Packaged solutions: Spidflow® Pack

Veolia has developed a full range of packaged and standardized Spidflow® Pack compact units for surface water treatment and wastewater refining.

Depending on the configuration and application of the Spidflow® Pack, the unit treatment capacity can reach from 100 to 643 $m^3 \cdot h^{-1}$.

This range of products is marketed by OTV – Veolia Water Technologies. The skids' shapes and their dimensions (1.5–3.5 m wide, arranged in rectangular tanks with a total height not exceeding 3.7 m) make them easily transportable by land and sea. The internal equipment is entirely supplied by OTV-VWT (mixers, surface scraper, nozzles, Turbomix, etc.).

The Spidflow® Pack comprises three reaction zones: a flocculation tank, a contact tank for the injection of supersaturated water and a separation tank (flotation zone). Flocculated water enters the contact zone and is then poured from above into the flotation zone.

Optimal flocculation is ensured thanks to a removable Turbomix' which is integrated into the flocculation tank with a specially adapted impeller. In case of need, a drain is provided at the structure's bottom to empty the flocculation tank. The skid is equipped for injecting reagents (coagulant and polymer).

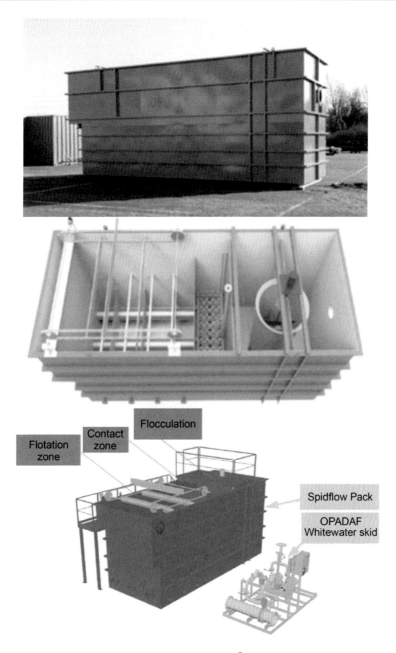

Figure 6.54. *Packaged solution for Spidflow® Pack. For a color version of this figure, see www.iste.co.uk/gaid/watertreatment1.zip*

Model	Maximum flow	Length	Width	Height/width ratio	Minimum flocculation time	Minimum injection time	Flotation surface	Maximum velocity
	Q (m³/h)	L (m)	l (m)	H/h (m)	$Tfloc$ (min)	$Tinj$ (min)	S (m²)	V (m/h)
SFP 400-F	135	6.1	1.5	3.65/3.45	5.1	1.2	3.8	40
SFP 500-F	218	7.2	2.0	3.65/3.45	5.1	1.1	6.0	40
SFP 600-F	318	8.3	2.5	3.65/3.45	5.1	1.0	8.8	40
SFP 700-F	434	9.4	3.0	3.65/3.45	5.1	1.0	12.0	40
SFP 800-F	643	11.6	3.5	3.65/3.45	5.1	0.9	17.5	40

Note: The flotation modeling was designed by Nicolas Roux (VERI-VEOLIA) and his team for the development of Spidflow®.

Table 6.19. Spidflow® Pack flow range and size

A perforated floor is set up in the contact zone and welded between the two tank walls. This floor comprises a ramp network equipped with supersaturated water injection nozzles. All these arrangements ensure a homogeneous distribution of supersaturated water and the best contact conditions with flocculated water. Two guides (one on each side of the tank) and a fixing system are arranged in the lower and lateral areas to facilitate the equipment's installation, withdrawal and technical assistance.

After solid–liquid separation, the water flows downward and is collected by a treated water collector via a network of perforated pipes placed on the floor of the flotation zone. A fraction of floated water is used for the preparation of supersaturated water. This is made from a gas saturation tank or directly using a specific pump.

Anti-spiral flow plates are designed to avoid hydraulic short circuits and a descent of the bubble bed, making it possible to obtain constant quality water. These plates are removable for easy maintenance.

Floating sludge is collected by a chain scraper and evacuated through a trough. This energy-efficient, free of oil leakage, low-maintenance scraper provides efficient sludge removal. A drainage for any decanted sludge is provided.

The skid metallic standard Spidflow® range is summarized in Table 6.19. The mirror settling velocities account for the recirculation flow, which can vary between 9 and 15% of the inlet flow.

6.8. References

Amato, T. and Wicks, J. (2009). Dissolved air flotation and potential clarified water quality based on computational fluid dynamics modelling. In *Proceedings of Water Quality Technology Conference*, AWWA, Seattle.

Amato, T., Edzwald, J.K., Tobiason, J.E., Dahlquist, J., Hedberg, T. (2001). An integrated approach to dissolved air flotation. *Water Science and Technology*, 43(8), 19–26.

Bernhardt, H. and Clasen, J. (1991). Flocculation of micro-organisms. *Journal Water SRT-Aqua*, 40(2), 76–87.

Betzer, N., Argaman, Y., Kott, Y. (1980). Effluent treatment and algae recovery by ozone induced flotation. *Water Research*, 14, 1003–1009.

Bratby, J. (1978). Aspects of sludge thickening by dissolved-air flotation. *Water Pollution Control*, 77(3), 421–432.

Bratby, J. and Marais, G. (1974). Dissolved air flotation. *Filtration et separation*, 614–624.

Briley, D. and Knappe, D. (2002). Optimizing ferric sulfate coagulation of algae with streaming current measurements. *J. Amer. Water Works Assoc.*, February, 80–90.

Chen, J. and Yeh, H. (2005). The mechanisms of potassium permanganate on algae removal. *Water Research*, 39, 4420–4428.

Chen, J. and Yeh, H. (2006). Comparison of the effects of ozone and permanganate preoxidation on algae flocculation. *Water Sci. Tech.: Water Supply*, 6(3), 79–88.

Chen, J., Yeh, H., Tseng, I. (2009). Effect of ozone and permanganate on algae coagulation removal – Pilot and bench scale tests. *Chemosphere*, 74, 840–846.

Dahlquist, J. and Göransson, K. (2004). Evolution of a high-rate dissolved air flotation process – From idea to full-scale application. In *Chemical Water and Wastewater Treatment*, Hahn, H., Hoffmann, E., Ødegaard, H. (eds). IWA Publishing, London.

De Rijk, S.E., Jaap H.J.M. aivan der, G., den Blanken, J.G. (1994). Bubble size in flotation thickening. *Water Research*, 28, 465–473.

Edzwald, J.K. (1995). Principles and applications of dissolved air flotation. *Water Science and Technology*, 31(3–4), 1–23.

Edzwald, J.K. (2007). Fundamentals of dissolved air flotation. *Journal of New England Water World Association*, 71(2), 89–112.

Edzwald, J. K. (2010).Dissolved air flotation and me. *Water Research*, 44, 2077–2106.

Edzwald, J.K. and Wingler, B.J. (1990). Chemical and physical aspects of dissolved-air-flotation for the removal of algae. *J. Water Suppl.: Res. & Technol. AQUA*, 39(1), 24–35.

Edzwald, J.K., Malley, J.P., Yu, C. (1990). A conceptual model for dissolved air flotation in water treatment. *Water Suppl.*, 8, 141–150.

Edzwald, J.K., Walsh, J.P., Kaminsky, G.S., Dunn, H.J. (1992). Flocculation and air requirements for dissolved air flotation. *J. Am. Water Works Assoc.*, 84(3), 92–100.

Edzwald, J.K., Tobiason, J.E., Amato, T., Maggi, L.J. (1999). Integrating high rate dissolved air flotation technology into plant design. *J. Am. Water Works Assoc.*, 91(12), 41–53.

Fleming, R.H. and Revelle, R. (1939). Physical processes in the ocean. Recent marine sediments. *Amer. Assoc. Petrol. Geol.*, 48–141.

Fukushi, K., Tambo, N., Matsui, Y. (1995). A kinetic model for dissolved air flotation in water and wastewater treatment. *Water Science and Technology*, 31(3–4), 37–48.

Fukushi, K., Matsui, Y., Tambo, N. (1998). Dissolved air flotation: Experiments and kinetic analysis. *J. Water Suppl.: Res. & Technol. AQUA*, 47(2), 76–86.

Gasnier, F., Gaïd, K., Bourdon, F. (2011). Spidflow® : la flottation à grande vitesse en eau potable. *L'eau, l'industrie, les nuisances*, 342, 37–42.

Haarhoff, J. and Edzwald, J.K. (2001). Modelling of floc-bubble aggregate rise rates in dissolved air flotation. *Water Science and Technology*, 43(8), 175–184.

Haarhoff, J. and Edzwald, J.K. (2004). Dissolved air flotation modelling: Insights and shortcomings. *J. Water Suppl.: Res. & Technol. AQUA*, 53(3), 127–150.

Han, M.Y. (2002). Modeling of DAF: The effect of particle and bubble characteristics. *J. Water Suppl.: Res. & Technol. AQUA*, 51(1), 27–34.

Han, M.Y., Park, Y.H., Yu, T.J. (2002). Development of a new method of measuring bubble size. In *2nd World Water Congress: Drinking Water Treatment*. IWA Publishing, London.

Hedberg, T., Dahlquist, J., Karlsson, D., Sorman, L.O. (1998). Development of an air removal system for dissolved air flotation. *Water Sci. Tech.*, 37(9), 81–88.

Heinanen, J., Jokela, P., Peltokangas, J. (1992). Experimental studies on the kinetics of flotation. *Chemical Water and Wastewater Treatment*, II, 247–262.

Henderson, R., Parsons, S., Jefferson, B. (2008).The impact of algae properties and pre-oxidation on solid-liquid separation of algae. *Water Research*, 42, 1827–1845.

Huang, Z., Legendre, D., Guiraud, P. (2011). A new experimental method for determining particle capture efficiency in flotation. *Chemical Engineering Science*, 66(5), 982–997.

Huang, Z., Legendre, D., Guiraud, P. (2012). Effect of interface contamination on particle-bubble collision. *Chemical Engineering Science*, 68(1), 1–18.

Legendre, D., Sarrot, V., Guiraud, P. (2009). On the particle inertia-free collision with a partially contaminated spherical bubble. *International Journal of Multiphase Flow*, 35, 163–170.

Leppinen, D.M. and Dalziel, S.B. (2004). Bubble size distribution in dissolved air flotation. *Journal of Water Supply Research & Technology – Aqua*, 53(8), 531–543.

Leppinen, D.M., Dalziel, S.B., Linden, P.F. (2000). Modelling the global efficiency of dissolved air flotation. In *Proceedings of the 4th International Conference Flotation in Water and Waste Water Treatment*, Helsinki.

Lundh, M., Jönsson, L., Dahlquist, J. (2000). Experimental studies of the fluid dynamics in the separation zone in dissolved air flotation. *Water Research*, 34(1), 21–30.

Lundh, M., Jönsson, L., Dahlquist, J. (2001). The flow structure in the separation zone of a DAF pilot plant and the relation with bubble concentration. *Water Science and Technology*, 43(8), 185–194.

Lundh, M., Jönsson, L., Dahlquist, J. (2002). Flow structures in a dissolved air flotation pilot tank and the influence on the separation of MBBR floc. *Water Science and Technology: Water Supply*, 2(2), 69–76.

Matsui, Y., Fukushi, K., Tambo, N. (1998). Modeling, simulation, and operational parameters of dissolved air flotation. *J. Water Suppl.: Res. & Technol. – AQUA*, 47(1), 9–20.

Montiel, A. and Welte, B. (1998). Preozonation coupled with flotation filtration: Successful removal of algae. *Water Science and Technology*, 37(2), 65–73.

Moruzzi, R.B. and Reali, M.A.P. (2010). Characterization of microbubble size distribution and flow configuration in DAF contact zone by a non-intrusive image analysis system and tracer tests. *Water Science and Technology*, 61(1), 253–262.

Martin-Ionesco, N. (1996). Une nouvelle combinaison des techniques d'oxydation par l'ozone et de la flotation. *L'eau, l'industrie et les nuisances*, 192, 32–34.

Ravasi, D., König, R., Principi, P., Perale, G., Demarta, A. (2019). Effect of powdered activated carbon as advanced step in wastewater treatments on antibiotic resistant microorganisms. *Current Pharmaceutical Biotechnology*, 20(63), 63–75.

Rodrigues, R.T. and Rubio, J. (2003). New basis for measuring the size distribution of bubbles. *Minerals Engineering*, 16, 757–765.

Roux, N., Daniel, O., Gasnier, F., Lemoine, C. (2011). Acquisition et analyse d'images en flottation à air dissous. Société Française de Génie de Procédés.

Roux, N., Vigneron-Larosa, N., Thouvenot, T., Gaïd, A. (2012). Modelling and operation of a Spidflow® plant. In *Conference IWA Flotation*, 29th–31st October. Columbia University, New York.

Rykaart, E.M. and Haarhoff, J. (1995). Behaviour of air injection nozzles in dissolved air flotation. *Water Science and Technology*, 31, 25–35.

Sarrot, V., Guiraud, P., Legendre, D. (2005). Determination of the collision frequency between bubbles and particles in flotation process. *Chemical Engineering Science*, 60(22), 6107–6117.

Sandbank, E. and Shelef, G. (1987). Harvesting of algae from high-rate ponds by flocculation-flotation. *Water Science and Technology*, 19(12), 257–263.

Sarrot, V., Huang, Z., Legendre, D., Guiraud, P. (2007). Experimental determination of particles capture efficiency in flotation. *Chemical Engineering Science*, 62, 7359–7369.

Schofield, T. (1994). Birmingham, Frankley water treatment plant redevelopment, IAWQ–IWSA–AWWA joint specialised conference. In *Flotation Processes in Water and Sludge Treatment*, Orlando.

Schofield, T., Perkins, R., Simms, J.S. (1991). Frankley water treatment works redevelopment, pilot scale studies. *JCIWEM*, 5(4), 370–385.

Shawcross, J., Tran, T., Nickols, D., Ashe, C. (1997). Pushing the envelope: Dissolved air flotation at ultra-high rate. In *CIWEM – International DAF Conference*, London.

Ta, C.T. and Beckley, J. (2001). A multiphase CFD model of DAF process. *Water Science and Technology*, 43(8), 153–157.

Tambo, N., Matsui, Y., Fukushi, K. (1986). A kinetic study of dissolved air flotation. In *World Congress of Chemical Engineering*, Tokyo.

Teixeira, M. and Rosa, M. (2006). Comparing dissolved air flotation and conventional sedimentation to remove cyanobacterial cells of Microcystis aeroginosa. *The Key Operating Conditions, Separation and Purification Technology*, 52, 84–94.

Valade, M.T., Edzwald, J.K., Tobiason, J.E., Dahlquist, J., Hedberg, T., Amato, T. (1996). Particle removal by flotation and filtration: Pretreatment effects. *J. Amer. Water Works Assoc.*, 88(12), 35–47.

Zabel, T.F. (1985). The advantages of dissolved air flotation for water treatment. *J. Amer. Water Works Assoc.*, 77(5), 42–46.

Index

Summary of Volume 2

Chapter 9. Adsorption on Activated Carbon

Index

Summary of Volume 3

Chapter 12. Biological Removal of Ammonia

12.1. The principle of biological nitrification
12.2. Design parameters
    12.2.1. Dissolved oxygen
    12.2.2. Filtration rate
    12.2.3. NH_4^+ concentration removed as a function of temperature (°C)
    12.2.4. Applicable volume load
    12.2.5. Contact time
    12.2.6. Material height
12.3. Factors limiting oxygen
    12.3.1. Mineral carbon
    12.3.2. pH
    12.3.3. Temperature
    12.3.4. Other elements
    12.3.5. Biological filter washing
12.4. Implementation
    12.4.1. Sand filtration
12.5. Biofilters (Biocarbon® process)
12.6. Water treatment stations
    12.6.1. Treatment stations with conventional sand, dual media or GAC filtration
    12.6.2. Treatment stations with Biocarbon® filters
12.7. References

Chapter 13. Nitrate Removal

13.1. Biological treatment
    13.1.1. Biochemical reactions
    13.1.2. Nitrite formation
    13.1.3. The bacteriological aspect
    13.1.4. Biofilter description (Biodenit® process)
    13.1.5. Water treatment stations including biological denitrification
    13.1.6. Factors affecting biological denitrification
    13.1.7. Design parameters: applied volumic load
    13.1.8. Design parameters: minimum contact time (tc min)
    13.1.9. Design parameters: height of Biodagen® material (m)
    13.1.10. Design parameters: material volume (m^3)
    13.1.11. Partial treatment
    13.1.12. Sludge production
    13.1.13. The reagents
    13.1.14. Biological denitrification implementation and exploitation

Summary of Volume 4

Chapter 17. Microfiltration and Ultrafiltration

Chapter 18. Nanofiltration and Reverse Osmosis

Index

Summary of Volume 5

Chapter 20. Calco-carbonic Equilibrium, Correction of Aggressivity and Remineralization

20.1. Characteristics of water leading to calco-carbonic equilibrium
    20.1.1. Chemical equilibria
    20.1.2. Aggressive water
    20.1.3. Scaling water
    20.1.4. Corrosive water
20.2. The equilibrium reactions of water's constituents
    20.2.1. Equilibrium pH
    20.2.2. Langelier equation
20.3. Hallopeau–Dubin diagram
20.4. Indicative criteria to determine the aggressivity or corrosivity of water
    20.4.1. Indicators of aggressivity: concrete pipelines
    20.4.2. Corrosivity indicators
20.5. The calco-carbonic equilibrium of water
    20.5.1. Water quality and regulations
    20.5.2. The correction of aggressivity
    20.5.3. Aggressivity correction treatments
20.6. Remineralization treatments
    20.6.1. Graphic method
    20.6.2. Processes for implementing remineralization: chemical reactions in tanks
20.7. Characteristics of the various reagents used
    20.7.1. Lime
    20.7.2. "Micronized" lime
    20.7.3. Caustic soda

Chapter 21. Disinfection

Chapter 22. Disinfection By-products

22.1. General aspects
22.2. Reaction by-products
22.3. Formation and evolution of chlorination by-products
22.4. Kinetics and formation mechanisms
    22.4.1. Formation kinetics
    22.4.2. Mechanisms
    22.4.3. Chlorination of HS
    22.4.4. Chlorination of carboxylic acids
    22.4.5. Factors influencing the formation of DBPs
22.5. Regulations
22.6. Predictive models of CBPs
22.7. Removal of THMs and HAAs
    22.7.1. Aeration
    22.7.2. Activated carbon
    22.7.3. Biofiltration
    22.7.4. High-pressure membranes
22.8. The case of nitrosamines and NDMA
    22.8.1. Nitrosation mechanism with HOCl
22.9. Oxidation by-products related to chlorine dioxide
22.10. Ozonation by-products
22.11. Recommendations
22.12. References

Chapter 23. Sludge Treatment

23.1. Choosing a treatment chain
23.2. Characteristics of drinking water sludge
    23.2.1. The quantity of sludge produced
    23.2.2. Sludge concentration estimate at different stages
    of the chain
    23.2.3. Sludge quality: physical and chemical properties
23.3. Handling and storage: shovelable and stackable nature
23.4. Different classes of sludge
    23.4.1. Hydroxide sludge
    23.4.2. Softening sludge
    23.4.3. Metal species sludge treatment
    23.4.4. Biological sludge
    23.4.5. The case of mixed sludge

Chapter 24. The Treatment Chain: Conception and Design

Chapter 25. The Future of Water

Index

Printed and bound by CPI Group (UK) Ltd, Croydon, CR0 4YY

23/08/2023

08103670-0001